T0139837

Advances in Intelligent Systems and Computing

Volume 767

The series "Advances in Intelligent Systems and Computing" contains publications on theory, applications, and design methods of Intelligent Systems and Intelligent Computing. Virtually all disciplines such as engineering, natural sciences, computer and information science, ICT, economics, business, e-commerce, environment, healthcare, life science are covered. The list of topics spans all the areas of modern intelligent systems and computing such as: computational intelligence, soft computing including neural networks, fuzzy systems, evolutionary computing and the fusion of these paradigms, social intelligence, ambient intelligence, computational neuroscience, artificial life, virtual worlds and society, cognitive science and systems, Perception and Vision, DNA and immune based systems, self-organizing and adaptive systems, e-Learning and teaching, human-centered and human-centric computing, recommender systems, intelligent control, robotics and mechatronics including human-machine teaming, knowledge-based paradigms, learning paradigms, machine ethics, intelligent data analysis, knowledge management, intelligent agents, intelligent decision making and support, intelligent network security, trust management, interactive entertainment, Web intelligence and multimedia.

The publications within "Advances in Intelligent Systems and Computing" are primarily proceedings of important conferences, symposia and congresses. They cover significant recent developments in the field, both of a foundational and applicable character. An important characteristic feature of the series is the short publication time and world-wide distribution. This permits a rapid and broad dissemination of research results.

** Indexing: The books of this series are submitted to ISI Proceedings, EI-Compendex, DBLP, SCOPUS, Google Scholar and Springerlink **

More information about this series at http://www.springer.com/series/11156

Atilla Elçi · Pankaj Kumar Sa ·
Chirag N. Modi · Gustavo Olague ·
Manmath N. Sahoo · Sambit Bakshi

Editors

Smart Computing Paradigms: New Progresses and Challenges

Proceedings of ICACNI 2018, Volume 2

 Springer

Editors
Atilla Elçi
Faculty of Engineering
Aksaray University
Sağlık, Aksaray, Turkey

Chirag N. Modi
National Institute of Technology Goa
Goa, India

Manmath N. Sahoo
Department of Computer Science
and Engineering
National Institute of Technology Rourkela
Rourkela, Odisha, India

Pankaj Kumar Sa
Department of Computer Science
and Engineering
National Institute of Technology Rourkela
Rourkela, Odisha, India

Gustavo Olague
CICESE
Ensenada, Baja California, Mexico

Sambit Bakshi
Department of Computer Science
and Engineering
National Institute of Technology Rourkela
Rourkela, Odisha, India

ISSN 2194-5357   ISSN 2194-5365 (electronic)
Advances in Intelligent Systems and Computing
ISBN 978-981-13-9679-3   ISBN 978-981-13-9680-9 (eBook)
https://doi.org/10.1007/978-981-13-9680-9

This Springer imprint is published by the registered company Springer Nature Singapore Pte Ltd.
The registered company address is: 152 Beach Road, #21-01/04 Gateway East, Singapore 189721, Singapore

Preface

It is a sheer pleasure announcing receipt of an overwhelming response from academicians and researchers of reputed institutes and organizations of the country and abroad for joining us in the 6th International Conference on Advanced Computing, Networking, and Informatics (ICACNI 2018), which makes us feel that our endeavor is successful. The conference organized jointly by the Department of Computer Science and Engineering National Institute of Technology Silchar and Centre for Computer Vision & Pattern Recognition, National Institute of Technology Rourkela during June 04–06, 2018, certainly achieves a landmark toward bringing researchers, academicians, and practitioners in the same platform. We have received more than 400 research articles and very stringently have selected through peer review the best articles for presentation and publication. We, unfortunately, could not accommodate many promising works as we strove to ensure the highest quality and adhere to the recommendations of the expert reviewers. We are thankful to have the advice of dedicated academicians and experts from industry and the eminent academicians involved in providing technical comments and quality evaluation for organizing the conference with a planned schedule. We thank all people participating and submitting their works and having continued interest in our conference for the sixth year. The articles presented in the two volumes of the proceedings discuss the cutting-edge technologies and recent advances in the domain of the conference. We conclude with our heartiest thanks to everyone associated with the conference and seeking their support to organize the 7th ICACNI 2019 at the Indian Institute of Information Technology Kalyani, India, during December 20–21, 2019.

Sağlık, Turkey   Atilla Elçi
Rourkela, India   Pankaj Kumar Sa
Goa, India   Chirag N. Modi
Ensenada, Mexico   Gustavo Olague
Rourkela, India   Manmath N. Sahoo
Silchar, India   Sambit Bakshi
November 2019

Contents

About the Editors

Atilla Elçi is an Emeritus Full Professor and Chairman of the Department of Electrical and Electronics Engineering at Aksaray University, Turkey (August 2012–September 2017). He previously served as a Full Professor and Chairman of Computer and Educational Technology at Süleyman Demirel University, Isparta, Turkey (May 2011–June 2012). Dr. Elçi graduated with a degree in Electrical Engineering from METU, Turkey and completed his M.Sc. and Ph.D. in Computer Science at Purdue University, USA. He has 71 papers, 11 books, 10 conference publications and 4 talks to his credit. His research areas include semantic web technology; information security/assurance; semantic robotics; programming systems and languages; and semantics in education.

Pankaj Kumar Sa received his Ph.D. degree in Computer Science in 2010. He is currently serving as an Associate Professor in the Department of Computer Science and Engineering, National Institute of Technology Rourkela, India. His research interests include computer vision, biometrics, visual surveillance, and robotic perception. He has co-authored a number of research articles in various journals, conferences, and book chapters and has co-investigated research and development projects funded by SERB, DRDOPXE, DeitY, and ISRO. He has received several prestigious awards and honors for his contributions in academics and research.

Chirag N. Modi is an Assistant Professor, Department of Computer Science and Engineering, NIT Goa, Farmagudi. He received his B.E. (Computer Engineering) from BVM Engineering College, Gujarat, India and completed his M.Tech. and Ph.D. in Computer Engineering at SVNIT Surat, Gujarat. His research areas include cloud computing security; network security, cryptography and information security; and secured IoTs, data mining, and privacy preserving data mining. He has published several journal papers and book chapters.

Gustavo Olague received his Ph.D. in Computer Vision, Graphics and Robotics from the INPG (Institut Polytechnique de Grenoble) and INRIA (Institut National de Recherche en Informatique et Automatique) in France. He is currently a

Professor at the Department of Computer Science at CICESE (Centro de Investigación Científica y de Educación Superior de Ensenada) in Mexico, and the Director of its EvoVisión Research Team. He is also an Adjoint Professor of Engineering at UACH (Universidad Autonóma de Chihuahua). He has authored over 100 conference proceedings papers and journal articles, coedited two special issues in Pattern Recognition Letters and Evolutionary Computation, and served as co-chair of the Real-World Applications track at the main international evolutionary computing conference, GECCO (ACM SIGEVO Genetic and Evolutionary Computation Conference). Professor Olague has received numerous distinctions, among them the Talbert Abrams Award presented by the American Society for Photogrammetry and Remote Sensing (ASPRS); Best Paper awards at major conferences such as GECCO, EvoIASP (European Workshop on Evolutionary Computation in Image Analysis, Signal Processing and Pattern Recognition) and EvoHOT (European Workshop on Evolutionary Hardware Optimization); and twice the Bronze Medal at the Humies (GECCO award for Human-Competitive results produced by genetic and evolutionary computation). His main research interests are in evolutionary computing and computer vision.

Manmath N. Sahoo is an Assistant Professor at the Department of Computer Science and Engineering, National Institute of Technology Rourkela, India. His research interest areas are in fault tolerant systems, operating systems, distributed computing, and networking. He is a member of the IEEE, Computer Society of India and the Institutions of Engineers, India. He has published several papers in national and international journals.

Sambit Bakshi is currently with Centre for Computer Vision and Pattern Recognition of National Institute of Technology Rourkela, India. He also serves as an Assistant Professor in the Department of Computer Science & Engineering of the institute. He earned his Ph.D. degree in Computer Science & Engineering in 2015. His areas of interest include surveillance and biometric authentication. He currently serves as associate editor of International Journal of Biometrics (2013–), IEEE Access (2016–), Innovations in Systems and Software Engineering: A NASA Journal (2016–), Expert Systems (2017–), and PLOS One (2017–). He has served/is serving as guest editor for reputed journals like Multimedia Tools and Applications, IEEE Access, Innovations in Systems and Software Engineering: A NASA Journal, Computers and Electrical Engineering, IET Biometrics, and ACM/Springer MONET. He is serving as the vice-chair for IEEE Computational Intelligence Society Technical Committee for Intelligent Systems Applications for the year 2019. He received the prestigious Innovative Student Projects Award 2011 from Indian National Academy of Engineering (INAE) for his master's thesis. He has more than 50 publications in reputed journals, reports, and conferences.

Research on Optical Networks, Wireless Sensor Networks, VANETs, and MANETs

Parallelization of Data Buffering and Processing Mechanism in Mesh Wireless Sensor Network for IoT Applications

Monika Jain, Rahul Saxena, Siddharth Jaidka and Mayank Kumar Jhamb

Abstract IoT, being a field of great interest and importance for the coming generations, involves certain challenging and improving aspects for the IoT application developers and researchers to work upon. A wireless sensor mesh networking has emerged as an attractive option for wide range of low-power IoT applications. This paper shows that how the data can be stored, read and processed parallelly by the parent node in the cluster from multiple sensor nodes, thus reducing the response time drastically. The use of parallelized algorithm for the communication protocol optimized using OpenMP standards for multi-core architecture between the sensors and parent node enables multiple radio technologies to be used for an application which could not be more than one in case of serial processing. The proposed algorithm has been tested for a wireless network application measuring temperature and moisture concentrations using numerous sensors for which the response time is recorded to be less than 10 ms. The paper also discusses in detail the hardware configurations for the application tested along with the results throwing light on the parallel mechanism for buffering and processing the messages. Finally, the paper is concluded by claiming the edge of parallel algorithm-based routing protocol over the serial in the light of graphical results and analysis.

Keywords Parallel algorithm · Wireless · Sensor · Mesh network topology · OpenMP · IoT

1 Introduction

Today, wireless sensor networks have become an inevitable part of the modern-day computing environment especially in the field of communication and information exchange. These kinds of networks have found their existence and most importantly significance in variety of applications such as military surveillance and monitoring, home automation systems, mode of transport like cars, airplanes, etc., industrial

M. Jain (✉) · R. Saxena · S. Jaidka · M. K. Jhamb
School of Computing and Information Technology, Manipal University Jaipur, Jaipur, India
e-mail: monikalnct@gmail.com

© Springer Nature Singapore Pte Ltd. 2020
A. Elçi et al. (eds.), *Smart Computing Paradigms: New Progresses and Challenges*,
Advances in Intelligent Systems and Computing 767,
https://doi.org/10.1007/978-981-13-9680-9_1

application like automating the machinery, environment analysis, weather predictions and many more [1]. The most important application of wireless sensor networks is data collection and processing. Owing to need of having high technological and computing in today's modern-day network scenario, there is an increased interest on data collection in the wireless sensor networks [2]. Usually, large or thousand numbers of sensor nodes are used for the collection of data. Various methods can be used for deployment of these nodes. The sensor nodes can be stationary or mobile. The ultimate challenge in this is how fast the data can be processed and the results can be brought back to the user. Keeping in mind the high data availability and exchange of information over the network, the need of applications is the instant responsiveness to the real world. This demands for some special computing measures and technique to be involved in order to process and compute such a large data in no time. In this paper, we come up with a concept of parallel read and data processing in an IoT-based application developed to read in the weather and moisture concentrations of an area. The code for the application has been modified using the OpenMP standard for parallel processing utilizing the parallel processing power of multi-core architecture of modern machines.

The paper has been modelled and organized into sections. Section 1 gives the general introduction to wireless networks, IoT-based applications and need of parallel processing in data buffering and reading. Section 2 discusses the wireless sensor mesh network architecture in brief. Section 3 discusses the various components of the network architecture of Sect. 2 in detail. Section 4 discusses the applicability of parallel processing in reading and processing the data from sensor nodes parallelly in the network. Section 5 presents the experimental set-up and working mechanism along with the hardware components and devices involved. Section 6 explains the incorporation of parallel processing to the mechanism explained in Sect. 6. Section 7 gives the experimental evidences in the light of performance comparison and graphical results that how the efficiency improvement in terms of abrupt responsiveness can be increased using the parallel processing power of multiple CPUs. Finally, the paper sums up with a Conclusion stating parallel implementation to have an incremental speed up over serial approach with the increasing number of sensor nodes in the network.

2 Wireless Sensor Network Architecture Using Mesh Topology

Wireless mesh networks work as regular wireless networks, but with some convincing differences. Mesh networks does not provide a centralized infrastructure rather it to maintain a decentral network by making each sensor node performs double roles, as a user and a router of data traffic. This way, the network exists as a self-organized and self-managed organic entity capable of providing service to a large number of users. Mesh networks typically rely on wireless nodes rather than centralized access points to create a virtual wireless backbone. Network "sensor nodes" establish network links with surrounding nodes, enabling traffic to hop between nodes on numerous paths through the network. The wireless sensor network can be deployed using three structures—(1) layered according to hop count (2) clustered (3) using mesh topology. The basic architecture comprises three major things: sensor nodes in which they collect all the information using various sensors like temperature, humidity, pressure, sound, etc., and they may also perform computation on the collected data and forward it further in the network [3]. Base nodes: All the processed data is stored here. Cluster nodes: The network can be divided into clusters. Each cluster consists of a cluster node which handles data of various sensor nodes present within the cluster [4, 5]. A wireless mesh network requires total $(N*(N-1))/2$ connections where N is the no. of nodes in the wireless mesh network.

3 Wireless Sensor Mesh Network Preliminaries

A wireless sensor mesh network is a self-healing network. When one node can no longer operate, rest of the nodes can still function and communicate with one another, directly or through individual nodes by automatically switching the paths between the nodes. These paths can be self-formed and self-healed depending upon the state of wireless nodes in the network. If a node is active and functional, it can contribute in making a path with the help of another node. Thus, a wireless network can have many interconnections. Wireless mesh networks can work with different wireless technologies including 802.11, 802.16, 02.16, cellular technologies, Wi-Fi, Bluetooth, radio frequency, etc. In addition, need not be restricted to any one technology or protocol. The wireless mesh network can also have multiple radios operating at different frequencies to interconnect the nodes in the network. Sensor node: The sensor node collects all useful information, if required, processing the data can also be done on the sensor nodes itself before transmitting the data further in the network.

A. *Multiple Sensor Node*

The network consists of multiple sensor nodes which coordinate with one another, exchange useful network information with one another to keep the network up and running all the time.

B. *Parent/Base Node*

The collected data is sent over to the base node for storing and processing further. The parent node can also perform actions based on collected data, like turning on the lights automatically when motion is detected in a room with home automation system. The parent node is aware of the sensor node from where the data originally came from, so it can send the feedback back to the same node. There can also be multiple parent nodes with in the network. Each parent node is responsible for handling group of sensors, thus forming a semi-cluster-based mesh network (Fig. 1).

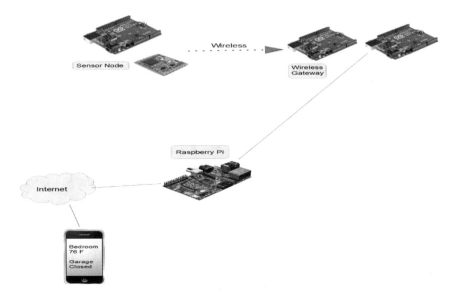

Fig. 1 Flow of data from sensors

C. *Routing Protocols*

There are various routing protocols [6] available for data collection in wireless sensor network: location-based protocols; data-centric protocols; hierarchical protocols; mobile protocols; multipath-based protocols; heterogeneity-based protocols; QoS-based protocols. The protocols can vary depending on the needs and the model used for the implementation of the wireless sensor mesh network.

D. *Wireless Technology*

There are various wireless technologies available for implementing the wireless sensor mesh network like 802.11, 802.15, 802.16, radio frequency, Bluetooth, etc.

4 Parallel Processing in Wireless Sensor Mesh Networks

Using an efficient dynamic protocol helps increasing the performance in the wireless sensor mesh network. This performance can further be increased by using parallel processing techniques. This can be done either by forming clusters of multiple parent nodes or by distributing the workload amongst the different cores of parent node for increased processing.

Processing of multiple buffers was first introduced by Intel in its I series processors [7]. Same technique can be applied to wireless sensor mesh networks also. There are many algorithms such as hashing or encryption, which are applied to stream of data buffers [8]. Since the data being processed is very large, there is an ever-increasing need for increasing the performance. Two basic ways to do this is to process multiple

Fig. 2 Sensor nodes sending the data to parent nodes with threads

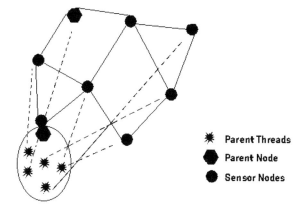

independent buffers in parallel or to process the buffers with SIMD instructions [9]. Processing in parallel fashion helps reducing data dependency limits.

In Fig. 2, the sensor nodes can send data to the parent node, which comprises multiple threads, i.e. multiple instances of parent node are present indirectly to process the data. More are the number of threads available on the parent side, faster the data can be processed, thus the interval set in the sensor nodes for sending information can be decreased drastically, for more accurate data logging and storing. Further benefits include improved response time in case of sending feedback to the respective sensor nodes for any action-related signals.

5 Experimental Set-up

For real-time implementation and analysis of the concept discussed in Section 4, we developed an application for real-time measurement of temperature and humidity concentrations of a given area using the sensor nodes. In this section, we will be discussing the hardware components and set-up for the application.

A. Hardware

- Parent node
  - SoC: Broadcom BCM2837
  - CPU: 1.2 GHz quad-core ARM Cortex A53
  - GPU: Broadcom Video Core IV @ 400 MHz
  - Memory: 1 GB LPDDR2-900 SDRAM
  - Sensor node
  - Microcontroller: Atmel328P
- RF module
  - Sensor node: NRF24l01
  - Parent node: NRF24L01+ with antenna

- Temperature and humidity sensor: DHT11
- Humidity: 20~90%
- Temperature: 0~+50 C
- Response time <10 ms.

Sensor nodes collect data and pair with the mesh network. The incoming messages are managed by a channel-based message subscription protocol called MQTT [10, 11]. The data is collected and parsed on the hub and pushed to cloud. The data is available to multiple devices both on local network and also over Internet enabling you to monitor your system anytime and anywhere.

B. Steps Involved
    There are four basic steps in the algorithm—collection, transmission, logging, reaction.

    Step 1—Collection
    - The data is collected on various sensor nodes individually. The data is in encrypted form.
    - After collection, the data is sent over mesh network to base hub.
    - Each sensor node is dynamically assigned a unique id so even if the sensor node gets disconnected from the network, it will repair the mesh network for continued transmission.
    - Each node is subscribed to an individual channel for separate handling by the threads.

    Step 2—Transmission
    - The data is transmitted in an encrypted format using various RF libraries [12].
    - The radio module has a max payload size limit of 32 bytes, being sent over a 2.4 GHz network at a speed of 2.4 MBps.
    - The message is queued on the board SOC and processed serially.

    Step 3—Logging
    The data is received on the hub, stored in SQLite database and pushed over cloud for over Internet usage on multiple devices.

    Step 4—Reaction
    The hub checks for the values and if found that any value exceeds the peak value, it sends the reaction message to that particular node.

6 Experimental Working

These algorithmic steps have been executed by employing OpenMP programming specifications over RF24 libraries by enabling the radio modules. An algorithmic pseudo-code has been represented below in order to show how these steps can be executed parallelly on each sensor.

```
Pseudo code: ParallelTemplogging
Begin
    #pragma omp parallel for num_threads (number of sensors) for (condition true)
        Update network.
        Check if the buffers are clear or not.
        If buffer is not clear, pick up the data from the buffer for processing.

        #pragma omp parallel
            for available network
                Initialize RF24 network header.
                Read header.
                Find the corresponding acknowledgment channel.
                Get temperature and humidity reading from packet.
                Store the data in database.
                Send acknowledgment packet to the respective sensor node.
            end for
    end for
End.
```

7 Experimental Results and Analysis

Following results have been evaluated over the set-up explained in Section 5. The parent node or base station where the data aggregates and processed to be sent to the end user uses the parallel processing power of the multi-core architecture that the machine has. The functions for reading and data processing tasks are split over the multiple CPU cores using the OpenMP [2, 13, 14] programming specification. For our analysis, the inputs are the temperature and humidity concentration values from three different sensor nodes, single base station. Using parallel processing, multiple instances handle the nodes in a parallel fashion for processing and reaction [15]. It is similar to have a personal thread for every sensor node. The following table presents a comparison in the execution time of serial and parallel implementation (Table 1).

On graphical plot above shows an incremental growth in the execution timing of the results produced as the number of sensor node increases. The execution timing is drastically cut down for the parallel implementation of the algorithm used which is required by real-time applications to provide instantaneous reactions and output to the end users (Fig. 3).

Table 1 Execution time comparison for serial and parallel versions

S. No.	Number of sensor nodes	Serial execution time (in seconds) T_s	Parallel execution time (in seconds) T_p	Speedup (T_s/T_p)
1	50	13.095	0.595	22.008
2	100	27.209	0.987	27.567
3	200	48.430	1.771	27.346
4	500	88.571	2.72	32.562
5	1000	159.638	4.521	35.310

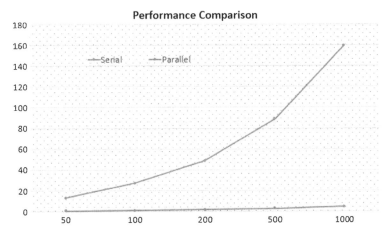

Fig. 3 Performance comparison graph for serial and parallel implementations; *x*-axis: number of sensor nodes; *y*-axis: execution time in seconds

Figure 4 shows the temperature log for the readings obtained from sensors read parallelly. Here a thread corresponds to each sensor node and is responsible for fetching the readings from that respective node. Parallel processing causes a difference when collection of data, forwarding and reaction is done by multiple threads [16, 17]. An interesting thing to note about the approach is that when parallel processing is applied, the increase in performance was drastic as the payloads were being queued on the SOCs. The messages are not forwarded until previous messages have successfully reached the destination and processed by it.

```
Data received from node 631 and processed by thread no. 605
Temperature : 24.000000
Humidity : 39.000000
.

.

.

Data received from node 287 and processed by thread no. 997
Temperature : 25.000000
Humidity : 45.000000
.

.

Data received from node 400 and processed by thread no. 372
Temperature : 30.000000
Humidity : 12.000000
.

.

.
```

Fig. 4 Output log for temperature (in degree centigrade) and humidity concentration (in percentage)

8   Conclusion and Scope

The paper presents a parallel data buffering and reading mechanism for applications involving large number of sensor nodes. With the enormously growing data, in order to achieve immediate spontaneity to the user query over web, we require some specialized techniques to present the results in real time. In this paper, we have designed an IoT-based application for reading temperature and humidity concentration data of a given place. By increasing the number of sensors in order to increase the coverage area slows down the rate at which the result is produced as each node interacts with every other node in the network as well as reports the results to the cluster head or parent node. This data retrieval and processing rate is enhanced by using the OpenMP programming model by mapping the data reading and buffering functions to parallel architecture of the machine having multiple CPUs. The paper also discusses in detail the experimental set-up along with the component description and connectivity. Graphical results and the comparison table in the previous section clearly indicate that the parallel implementation has a high-speedup gain over the serial implementation. As the number of nodes in the network increases, the performance gain also increases. Further investigation can be done adopting the same theory of multiple parent nodes for a cluster-based computing environment as single machine meets to a certain limitation at some instance.

The heavy utilization of the RF modules results in heating up of the modules in case of prolonged usage, which affects the performance and speedup a bit. Instead of software implementation of data encryption, if the SOC is designed in a way such that the encryption is done on board, the performance can be increased further.

References

1. F. Bendali et al., in *An Optimization Approach For Designing Wireless Sensor Networks*. New Technologies, Mobility and Security, 2008. NTMS '08. IEEE, 2008
2. R, Saxena, M. Jain, D. Singh, A. Kushwah, in *An Enhanced Parallel Version of RSA Public Key Crypto Based Algorithm Using OpenMP*. Proceedings of the 10th International Conference on Security of Information and Networks, ACM (Oct 2017), pp. 37–42
3. C. M. Nguyen et al., in *Wireless Sensor Nodes For Environmental Monitoring In Internet of Things. Microwave Symposium (IMS)*, 2015 IEEE MTT-S International. IEEE, 2015
4. O. Younis, M. Krunz, S. Ramasubramanian, Node clustering in wireless sensor networks: Recent developments and deployment challenges. IEEE Netw. **20**(3), 20–25 (2006)
5. K. Kumaravel, A. Marimuthu, in *An Optimal Mesh MESH Routing Topology Using Mesh In Wireless Sensor Networks*. 2014 International Conference on Green Computing Communication and Electrical Engineering (ICGCCEE), IEEE, 2014
6. S.K. Singh, M.P. Singh, D.K. Singh, Applications, classifications, and selections of energy-efficient routing protocols for wireless sensor networks. Int. J. Adv. Eng. Sci. Technol. (IJAEST) **1**(2), 85–95 (2010)
7. S. Akhter, J. Roberts, in *Multi-core programming*. (Hillsboro, 2006 Intel press), vol. 33
8. F. Rivera, M. Sanchez-Elez, M. Fernandez, N. Bagherzadeh, in *An Approach To Execute Conditional Branches Onto SIMD Multi-Context Reconfigurable Architectures*. 8th Euromicro Con-

ference on Digital System Design (DSD'05), 2005, pp. 396–402. https://doi.org/10.1109/dsd.
2005.14

9. K.-H. Chen et al, in *SIMD Architecture For Job Shop Scheduling Problem Solving*. ISCAS
   2001. The 2001 IEEE International Symposium on Circuits and Systems, vol. 4, IEEE, 2001

10. U. Hunkeler, H. L. Truong, A. Stanford-Clark, in *MQTT-S—A Publish/Subscribe Protocol for
    Wireless Sensor Networks*. 2008 3rd International Conference on Communication Systems
    Software and Middleware and Workshops, COMSWARE 2008. IEEE, 2008

11. D. Thangavel et al., in *Performance Evaluation of MQTT and CoAP Via A Common Middle-
    ware*. 2014 IEEE Ninth International Conference on Intelligent Sensors, Sensor Networks and
    Information Processing (ISSNIP), IEEE, 2014

12. J. A. Fisher, B. G. Brian, L. G. Jesionowski, *Selective Encryption Of Data Stored on Removable
    Media in an Automated Data Storage Library*, U.S. Patent No. 9,471,805, 18 Oct 2016

13. R. Saxena, M. Jain, D. P. Sharma, A. Mundra, in *A Review Of Load Flow and Network Recon-
    figuration Techniques With Their Enhancement For Radial Distribution Network*. 2016 Fourth
    International Conference on Parallel, Distributed and Grid Computing (PDGC), (Waknaghat,
    2016), pp. 569–574

14. R. Saxena, M. Jain, D. P. Sharma, GPU-based parallelization of topological sorting, in *Proceed-
    ings of First International Conference on Smart System, Innovations and Computing. Smart
    Innovation, Systems and Technologies*, vol. 79, ed. by A. Somani, S. Srivastava, A. Mundra, S.
    Rawat (Springer, Singapore, 2018)

15. M. Jain, R. Saxena, Parallelization of video summarization over multi-core processors. Int. J.
    Pure Appl. Math. **118**(9), 571–584 (2018). ISSN 1311-8080 (printed version); ISSN 1314-3395
    (on-line version)

16. M. Jain, R. Saxena, V. Agarwal, A. Srivastava, An OpenMP-Based Algorithmic Optimization
    for Congestion Control of Network Traffic, in Information and Decision Sciences. Advances
    in Intelligent Systems and Computing, vol. 701, ed. by S. Satapathy, J. Tavares, V. Bhateja, J.
    Mohanty (Springer, Singapore, 2018)

17. R. Saxena, M. Jain, S. Bhadri, S. Khemka, in *Parallelizing GA Based Heuristic Approach for
    TSP over CUDA and OPENMP*, 2017 International Conference on Advances in Computing,
    Communications and Informatics (ICACCI), (Udupi, 2017), pp. 1934–1940. https://doi.org/
    10.1109/icacci.2017.8126128

Modular Rook (MR): A Modular Data Center Network Architecture Based on Rook's Graph

Riju Das and Nabajyoti Medhi

Abstract Data center networking is becoming more challenging day by day in to-day's world of cloud computing. Due to fast technical innovation and vendor diversity, heterogeneity widely exists among data center networks (DCN). Increasing service requirements by industries or organizations requires expansion of the data center network. In the construction of the mega data centers, shipping container-based modular networks are used as basic building blocks. In this paper, we propose Modular Rook (MR) Network and a new modular data center (MDC) network design using Rook's graph for inter-container network. Modular Rook (MR) combines the advantages of both structured inter-container network and heterogeneous container networks. Modular Rook (MR) is found to possess brilliant network properties such as low network diameter and high bisection bandwidth with a fewer number of switches and links as compared to other DCN topologies.

Keywords Data center network (DCN) · Equal-cost multi-path (ECMP) routing · Modular data center (MDC) network · Modular Rook (MR)

1 Introduction

Data center networks (DCNs) are expanding exponentially to adapt the increasing application demands and user requirements. In order to design an interconnection network, it is essential to have a comprehensive understanding about properties and limitations of the network. These properties and limitations are characterized by the topology of the network. Since a topology sets constraints and costs, it plays a

R. Das (✉)
Department of Computer Science & Engineering, College of Technology
and Engineering, Udaipur 313001, Rajasthan, India
e-mail: rijjudas@gmail.com

N. Medhi
Department of Computer Science & Engineering,
Tezpur University, Tezpur 784028, Assam, India

© Springer Nature Singapore Pte Ltd. 2020
A. Elçi et al. (eds.), *Smart Computing Paradigms: New Progresses and Challenges*,
Advances in Intelligent Systems and Computing 767,
https://doi.org/10.1007/978-981-13-9680-9_2

critical role in all interconnection networks. Typical tree-type network architecture is a hierarchy of switches and routers. With the increase of network size, the hierarchy of these elements also increases which requires more powerful, deeper, and more extensive switches and routers.

The trend of modularizing a network architecture is increasing remarkably in today's world. Containerized or modularized data center is considered as the primary unit for the construction of modern data center (MDC) network [1]. An MDC container network contains a large number of switches and servers in a single 20–40 feet shipping container. MDC architecture provides various advantages [2–4] like less cabling complexity, faster deployment speed, scalability in terms of network capacity, power, and cost efficiency. Heterogeneity in the network is very obvious because of two reasons—(i) fast hardware innovation, due to which the owners of data centers purchase network hardware frequently to fit the increasing services; (ii) the container purchased from different sellers may use different topologies for connection of network elements within the container based upon the requirements of the customer.

Looking at the advantages of MDC network and addressing the issues of heterogeneity, we propose a new MDC architecture named Modular Rook (MR) which adopts Rook's Graph for inter-container network.

2 Related Works

In this section, we discuss about some of the notable existing data center network topologies. A DCN topology can be categorized as either switch-centric or server-centric topology. A switch-centric topology uses switches for forwarding packets, and a server-centric topology uses servers for the same. A typical tree-based switch-centric DCN topology, such as Clos topology [5, 6], consists of core layer and top-of-rack (ToR) layer switches. A Fat tree [7] is a special instance of Clos network, it has a three-layer tree structure with multiple roots, and these layers are named as core layer, aggregation layer, and edge layer. Server-centric topologies shift the workload from switches or routers to servers, that, is servers or hosts participate in routing and packet forwarding, while low commodity switches interconnect a fixed number of servers. BCube [8] is server-centric structure that contains two types of elements, (i) servers or hosts having multiple ports, (ii) switches which link to a fixed number of hosts. BCube topology is a recursively defined network which provides rich all-to-all connectivity among hosts. MDCube [9] is a modular DCN architecture built by interconnecting BCube-based container networks. It is built using commodity-off-the-shelf (COTS) switches where high-speed links of the switches are used for inter-container connections. Due to the use of COTS switches, MDCube proves to be a cost-effective design for modular DCNs.

3 Modular Rook Architecture: Background and Construction

In Modular Rook architecture, we use Rook's graph as the interconnection network of the modules or containers. Each module can be a different topology having a low diameter.

3.1 Background

Rook's Graph: Rook's graph is a concept from graph theory, and it is a graph which represents moves of a rook in a chess game. Rook's graph is a highly symmetric perfect graph. Figure 1 shows a 4×4 Rook's graph.

Vertices of a $n \times m$ Rook's graph is denoted by (i, j), where $1 \leq i \leq n$ and $1 \leq j \leq m$. Any 2 vertices (a_1, b_1) and (a_2, b_2) of a Rook's graph is called neighbors or adjacent only when either $a_1 = a_2$ or $b_1 = b_2$. Total number of vertices and total number of edges of a Rook's graph are $n \cdot m$ and $n \cdot m \cdot (n + m)/2 - n \cdot m$, respectively. Diameter of a Rook's graph is always two.

Fig. 1 A 4×4 Rook's graph

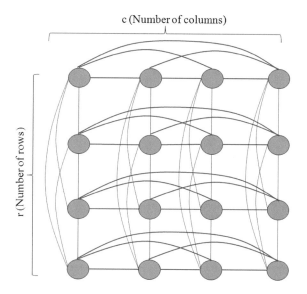

3.2 Modular Rook Network Construction

Modular Rook (MR) architecture is mainly designed for modular data center network. In MR, the Rook's graph is used for interconnection of the containers. Each node of the outer Rook's graph is a container containing servers and switches. Each container can have different topologies for the connection of its switches and servers. Modular Rook (MR) Network is designed to interconnect a large number of heterogeneous containers by using the structure of Rook's graph. In order to reduce the diameter of the overall structure, low diameter graphs are preferred for the container network. For intra-container connections, almost the same capacity networks are taken so that 1:1 over-subscription can be maintained at all the containers. MR can be constructed using commonly available COTS switches where high speed 10 Gbps links are used for inter-container connections and 1 Gbps links are used for switch to host connections.

Modular Rook (MR) can be denoted as $G_{(r \times c)}$, where r and c are the number of rows and columns of the outer Rook's graph, respectively. And a container of MR is denoted by C_{xy}, where $1 \leq x \leq r$ and $1 \leq y \leq c$; i.e., C_{xy} is the container positioned in the yth column of xth row. A switch inside a container C_{xy} is represented by a pair of container id and its id inside that container, i.e., $\{C_{xy}, S_{i\_xy}\}$.

Algorithm 1 Procedure for interconnection of containers of Modular Rook (MR)

1: **procedure** CONSTRUCTMODULARROOK(r, c)
                                                   ▷ Row-wise connections of containers
2:     **for** x=0 to r-1 **do**
3:       **for** y=0 to c-1 **do**
4:         **for** z=y+1 to c-1 **do**
5:           **Connect** C_{xy} **and** C_{xz}
6:         **end for**
7:       **end for**
8:     **end for**
                                                   ▷ Column-wise connections of containers
9:     **for** x=0 to c-1 **do**
10:       **for** y=0 to r-1 **do**
11:         **for** z=y+1 to r-1 **do**
12:           **Connect** C_{xy} **and** C_{xz}
13:         **end for**
14:       **end for**
15:     **end for**
16: **end procedure**

Algorithm 1 shows the procedure to connect the containers in a row-wise and column-wise manner to build the Rook's graph structure. According to Algorithm 1, a container C_{xy} has direct links to all other containers those are there in the xth row and yth column. MR can be expanded either row-wise or column-wise thus forming the nearest possible chessboard-type rectangular formation.

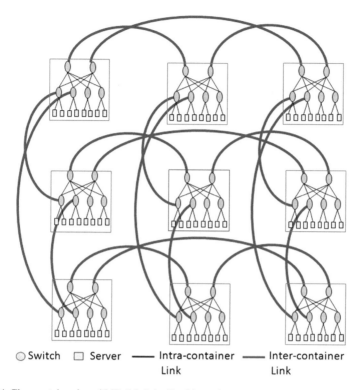

Fig. 2 A Clos container-based MR (Modular Rook) topology

Figure 2 shows a homogeneous Clos container-based (3×3) MR topology consisting of nine containers.

4 Properties of Modular Rook (MR)

Some of the basic properties of Modular Rook (MR) Network are given below.

Theorem 1 *If the total number of links in $(r \times c)$ Modular Rook Network $G_{r \times c}$ is denoted as $L_{(r \times c)}$, then we can say that, $L_{r \times c} = \frac{r \cdot c (r+c)}{2} - r \cdot c + r \cdot c \cdot n_l$, where n_l is the average number of intra-container links in each container C_{xy} of $G_{r \times c}$.*

Proof Based on the structure of a Rook's graph, for a $(r \times c)$ Modular Rook network $G_{r \times c}$, we will have a total $r \times c$ number of containers.

Suppose a container is denoted as C_{xy}, where $x \in [0, r-1]$ and $y \in [0, c-1]$.

According to structure of Rook's graph, a container C_{xy} has direct links to all other containers those are there in the xth row and yth column (as shown in Fig. 1); i.e., each container has total $(r-1) + (c-1) = (r+c-2)$ number of inter-container links.

Therefore, the number of total inter-container links is equal to $r \cdot c \cdot (r + c - 2) = r \cdot c \cdot (r + c) - 2 \cdot r \cdot c$

Now suppose, each container has n_l number of links in $G_{r \times c}$. So, the btotal number of intra-container links is equal to $r \cdot c \cdot n_l$.

Thus, the total number of links of $(r \times c)$ Modular Rook $G_{r \times c}$ is equal to intra-container links + inter-container links, i.e.,

$$L_{r \times c} = r \cdot c(r + c) - 2 \cdot r \cdot c + r \cdot c \cdot n_l$$

Theorem 2 *If the total number of servers in $(r \times c)$ Modular Rook(MR) is denoted as $N_{(r \times c)}$, then we can say that $N_{r \times c} = H \cdot r \cdot c \cdot (r + c - 2)$ where H is the bandwidth of the high-capacity switch port.*

Proof Based on the structure of a Rook's graph, for a $(r \times c)$ MR graph $G_{r \times c}$, we will have total $r \cdot c$ number of containers.

Suppose a container is denoted as C_{xy}, where $x \in [0, r - 1]$ and $y \in [0, c - 1]$.

According to structure of Rook's graph, a container C_{xy} has direct links to all other containers those are there in the xth row and yth column(as shown in Fig. 1); i.e., each container has total $(r - 1) + (c - 1) = (r + c - 2)$ number of inter-container links.

Suppose each inter-container link is of H Gbps capacity, therefore, a container can have a maximum $H \cdot (r + c - 2)$ number of servers to achieve 1:1 over-subscription ratio between the incoming and outgoing traffic to and from the container.

Thus, for $r \cdot c$ number of containers of a $G_{r \times c}$ Modular Rook, we can say that total number of servers is,

$$N_{r \times c} = H \cdot r \cdot c \cdot (r + c - 2)$$

Theorem 3 *Suppose, a Modular Rook (MR) $G_{r \times c}$ has homogeneous containers; i.e., each container C_{xy} has the same topology for intra-container connections. Then, we can say that the diameter of $G_{r \times c}$ is, $d_{r \times c} = 3d + 2$, where d is the diameter of the intra-container topology.*

Proof The diameter of a Rook's graph is always 2 (according to its structure). Therefore in MR, the diameter of inter-container connections is 2.

From Fig. 3, we can see that container C_{00} and container C_{33} are one of the far away nodes in $G_{r \times c}$. To reach C_{33} from C_{00}, we have to go through minimum one node, i.e., via two links. (One of the minimum path between C_{00} and C_{33} is shown in Fig. 3 by the red line).

We assume that each container has d-diameter graph for its intra-container connection. So, the diameter of Modular Rook $G_{r \times c}$ is,

$$d_{r \times c} = d + d + d + 2 = 3d + 2$$

Fig. 3 Path between
container 00 and container
33 of G_r is showing by the
darker red line

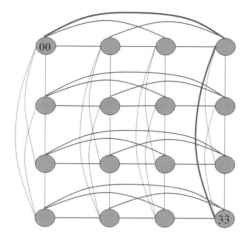

Theorem 4 *If we denote the bisection bandwidth of an even dimensional homogeneous $(r \times c)$ Modular Rook (MR) as $BiW_{r \times c}$, then we can say that $BiW_{r \times c} = H \cdot \frac{r \cdot c^2}{4}$, where number of rows ($r$) and columns ($c$) are even numbers and H is the bandwidth of the high-capacity switch port.*

Proof Bisection bandwidth of a network is the bandwidth available between two parts of the network, when we divide the network into two equal parts. Bisection of the network has to be done in such a way that bandwidth between these two parts is minimum. Bisection bandwidth accounts for the bottleneck bandwidth of the network.

Based on the structure of a Rook's graph, for a $(r \times c)$ MR, we will have total $(r \cdot c)$ number of containers.

A container C_{xy} has direct links to all the other containers in its same (xth) row and same (yth) column.

For even number of rows (r) and columns (c), we can divide the network into two equal parts row-wise or column-wise by the cut C, each having $\frac{r \cdot c}{2}$ number of containers.

For each row in $G_{r \times c}$, we have c number of containers. If we divide the network in row-wise manner then, for each row half of the containers, i.e., $\frac{c}{2}$ containers have direct links to another half of the containers in that row. Therefore we can say that, for each container, the row-wise cut C will have to go through $\frac{c}{2}$ number of its links. Thus, for Modular Rook (MR), $G_{r \times c}$, the cut C will cut total $\frac{c}{2} \times \frac{r \cdot c}{2} = \frac{r \cdot c^2}{4}$ number of links. The capacity of each inter-container link is H Gbps. Therefore, we can say that the bisection bandwidth of Modular Rook $G_{r \times c}$ is,

$$BiW_{r \times c} = H \cdot \frac{r \cdot c^2}{4}$$

5   Routing in Modular Rook (MR)

For the efficient communication of packets in a network, a fast and simple routing algorithm should be used, which ensures that the forwarding of packets from source to destination is done via the shortest path. Equal-cost multi-path routing (ECMP) is an efficient multi-path routing strategy that splits the traffic over multiple equal-cost links for load balancing. By load balancing the traffic over multiple paths, ECMP offers considerable increases in bandwidth of the network. In Modular Rook, for inter-container routing, ECMP is used and routing for intra-container depends on the topology of the container.

The diameter of a Modular Rook graph is $3d + 2$ (as explained in Theorem 3). And the diameter of the outer Rook's graph is 2. Any pair of containers $\{C_{x_1y_1}, C_{x_2y_2}\}$ of Modular Rook (MR) $G_{r \times c}$ has at least one single shortest path, having path length ≤ 2. In general, a pair of containers $\{C_{x_1y_1}, C_{x_2y_2}\}$ has the shortest path of length 1 between them if either $x_1 = x_2$ or $y_1 = y_2$, i.e., both are in same row and same column, and it has two shortest paths of length 2 if both are in different rows and columns. Thus, MR has several available shortest paths among the pairs of containers. Thus, the availability of multiple parallel equal-cost paths makes ECMP (equal-cost multi-path) a suitable routing algorithm in MR for routing of packets among the containers.

Figure 4 shows the number of shortest paths available between any pair of containers in Modular Rook Network. Figure 4a shows two parallel edges and node disjoint shortest paths (highlighted by red and green lines) between the container pair $\{C_{00}, C_{33}\}$ belonging to different rows and columns. The container pair of $\{C_{00}, C_{33}\}$ two shortest paths of length is equal to 2. Figure 4b shows a single shortest path (highlighted by the red lines) between the container pairs $\{C_{00}, C_{02}\}$ and $\{C_{00}, C_{30}\}$.

Due to its modular structure, DCN fabrics such as OSCAR [10] can be efficiently deployed for routing in MR since OSCAR easily fits into modular DCN topologies.

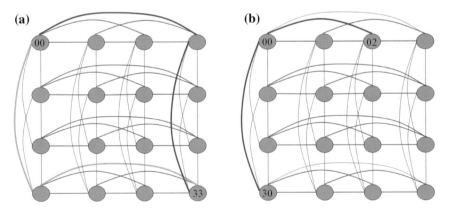

Fig. 4 Shortest path between a pair of containers **a** belong to different rows or columns and **b** belonging to same row or column

6 Comparison of DCN Topologies

In this section, the properties of different data center network topologies are compared. A qualitative and quantitative comparison [11] is done of different switch- and server-centric topologies which are shown in Table 2. A comparison is performed by taking various important parameters like the number of ports (n) and the number of levels (k). Descriptions of the symbols used in equations of different properties are given in Table 1.

6.1 Comparison of Network Elements of Modular Rook with Other Topologies

The complexity of a network depends on the number of physical elements connected to that network like the number of switches and links. Reducing and minimizing these network elements is a key consideration for data center network deployment. Figure 5 shows the graphs for variations of the number of switches and number of links respect to a number of servers for different topologies. Figure 5a shows number of switches versus number of servers. From the Fig. 5a, we observe that the number of switches required in Modular Rook is less than MDCube, BCube, and Fat Tree. This guarantees the reduction of the network cost due to switches. Similarly from the Fig. 5b, we find that the number of links required to interconnect MR is less than that of all the other topologies. This ensures low cabling complexity for MR.

Table 1 Symbol description

Symbol	Description
N	Total no. of servers in a topology
n	No. of ports of a switch
D	No. of dimensions
H	Bandwidth of a high-capacity switch link
l	No. of layers or levels
h	Height of the tree
e_0	Degree of switches of a level-0 graph
s_0	No. of nodes of a level-0 graph
r, c	No. of rows and columns of MR (Modular Rook) network respectively
n_l	No. of links in each container of MR (Modular Rook) network

Table 2 Quantitative and qualitative comparison of different topologies

Topology	No. of servers	No. of switches	No. of links	Diameter	Bisection bandwidth
Basic tree [12]	$(n-1)^3$	$(\frac{n^2+n+1}{n^3})N$	$(\frac{n}{n-1})(N-1)$	$2\log_{n-1} N$	1
Fat tree [7]	$\frac{n^3}{4}$	$\frac{5N}{n}$	$N\log_{\frac{n}{2}}\frac{N}{2}$	6	$\frac{N}{2}$
Clos network [7]	$\frac{n^2}{4}n$	$\frac{3}{2}n+\frac{n^2}{4}$	$N+\frac{4N}{n}$	6	$\frac{N}{2}$
BCube [8]	n^{l+1}	$\frac{N}{n}\log_n N$	$\log_n N$	$\log_n N$	$\frac{N}{4\log_n N}$
Ficonn [13]	$\geqslant 2^{l+2}\left(\frac{s_0}{4}\right)^{2^l}$, for $s_0 > 4$	$\frac{N}{n}$	$2N$	$2^{l+1}-1$	$\frac{N}{4\times 2^k}$
Hyper BCube [14]	n^{2l-1}	ln^{2l-2}	ln^{2l-1}	$\leqslant 3l-2$	$\frac{N}{n^{2l-1}-1}{4}$; $l \geqslant 2$
MDCube [9]	$\frac{n(l+1)}{D}^D n^{l+1}$	$\frac{N}{n}(l+1)$	$N(l+1)+\frac{N}{2n}(l+1)$	$4l+3+D(2l+3)$	$\frac{N}{2}$
Modular Rook	$Hrc(r+c-2)$	rcn_c	$rc(r+c)-2rc+rcn_l$	$3d+2$	$H\frac{rc^2}{4}$

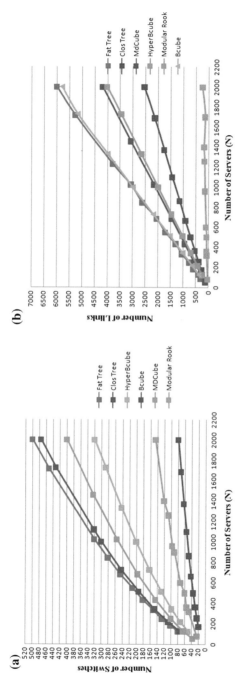

Fig. 5 Comparison of **a** number of switches **b** number of links required for the same number of servers of different topologies

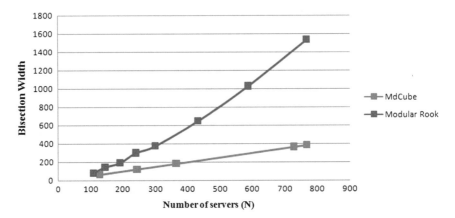

Fig. 6 Comparison of bisection bandwidth with respect to the number of servers of Modular Rook and MDCube

6.2 Comparison of Bisection Bandwidth of Modular Rook with Other Topologies

Figure 6 shows the variation of bisection bandwidth with respect to the number of servers for the MDC architectures—MDCube and Modular Rook (MR). It clearly shows the bandwidth advantage of MR as compared to that of MDCube.

7 Conclusion

In this paper, we propose a new modular data center network architecture named Modular Rook (MR) which uses Rook's graph for interconnecting heterogeneous container networks. Construction of MR and discussion on some of its properties like the number of physical elements (switch, server, and link), diameter, and bisection bandwidth are provided. A comparison of the topological properties of MR with some other existing popular DCN topologies shows a significant advantage of MR over others. This makes MR a suitable and easily deployable MDC network topology.

References

1. K.V. Vishwanath, A. Greenberg, D.A. Reed, Modular data centers: how to design them? in *Proceedings of the 1st ACM Workshop on Large-Scale System and Application Performance* (ACM, 2009), pp. 3–10
2. K. Wu, J. Xiao, L.M. Ni, Rethinking the architecture design of data center networks. Front. Comput. Sci. 1–8 (2012)

3. J. Bell, Top 6 benefits of modular data centers, https://www.colocationamerica.com/blog/top-6-modular-data-center-benefits

4. F. Yao, J. Wu, G. Venkataramani, S. Subramaniam, A comparative analysis of data center network architectures, in *2014 IEEE International Conference on Communications (ICC)* (IEEE, 2014), pp. 3106–3111

5. Introduction to Clos networks, https://web.stanford.edu/class/ee384y/Handouts/clos_networks.pdf

6. H. Weatherspoon, Data center network topologies: FatTree, https://www.cs.cornell.edu/courses/cs5413/2014fa/lectures/08-fattree.pdf

7. M. Al-Fares, A. Loukissas, A. Vahdat, A scalable, commodity data center network architecture, in *ACM SIGCOMM Computer Communication Review*, vol. 38, no. 4 (ACM, 2008), pp. 63–74

8. C. Guo, G. Lu, D. Li, H. Wu, X. Zhang, Y. Shi, C. Tian, Y. Zhang, S. Lu, Bcube: a high performance, server-centric network architecture for modular data centers, in *ACM SIGCOMM Computer Communication Review*, vol. 39, no. 4 (2009), pp. 63–74

9. H. Wu, G. Lu, D. Li, C. Guo, Y. Zhang, MDCube: a high performance network structure for modular data center interconnection, in *Proceedings of the 5th International Conference on Emerging Networking Experiments and Technologies* (ACM, 2009), pp. 25–36

10. N. Medhi, D. Saikia, Openflow-based scalable routing with hybrid addressing in data center networks. IEEE Commun. Lett. **21**, 1047–1050 (2017)

11. T. Chen, X. Gao, G. Chen, The features, hardware, and architectures of data center networks: a survey. J. Parallel Distrib. Comput. **96**, 45–74 (2016)

12. C.E. Leiserson, Fat-trees: universal networks for hardware-efficient supercomputing. IEEE Trans. Comput. **100**(10), 892–901 (1985)

13. D. Li, C. Guo, H. Wu, K. Tan, Y. Zhang, S. Lu, FiConn: using backup port for server interconnection in data centers, in *IEEE Infocom 2009* (IEEE, 2009), pp. 2276–2285

14. D. Lin, Y. Liu, M. Hamdi, J. Muppala, Hyper-BCube: A scalable data center network, in *2012 IEEE International Conference on Communications (ICC)* (IEEE, 2012), pp. 2918–2923

A Survey on Various Handoff Methods in Mobile Ad Hoc Network Environment

Libin Thomas, J. Sandeep, Bhargavi Goswami and Joy Paulose

Abstract Communication has never been the same since the advent of cellular phones and numerous applications with different functionalities seem to crop up on a daily basis. Various applications seem to crop up on a daily basis. Ad hoc networks were developed with the intent of creating networks made up of interconnected nodes, on-the-go. Ad hoc networks have numerous applications, the most popular being vehicular ad hoc networks (VANETs). In VANETs, moving vehicles are considered to be the mobile nodes and mobile vehicular nodes move at high speeds. Mobility of the nodes makes it difficult to maintain stable communication links between the nodes and the access points. A process known as handoff is used to bridge this gap and is considered to be one of the solutions for unstable communication links over larger distances. Handoff can usually be seen when the nodes are mobile and start to move away from the access points. This paper discusses and compares various handoff methods that were proposed by various researchers with an intent to increase positive attributes while negating the rest of the components that do not support in increasing the efficiency of the handoff process.

Keywords Ad hoc networks · Handoff · VANET · Switching · Routing · Access points · Cross layer · Vertical handoffs · Cost-based methods · Cognitive radio · Wireless communications · Seamless · Horizontal handoff

L. Thomas (✉) · J. Sandeep · B. Goswami · J. Paulose
Department of Computer Science, Christ (Deemed to be University), Hosur Road, Bangalore, Karnataka 560029, India
e-mail: thomaslibin82@gmail.com

J. Sandeep
e-mail: sandeep.j@christuniversity.in

B. Goswami
e-mail: bhargavi.goswami@christuniversity.in

J. Paulose
e-mail: joy.paulose@christuniversity.in

© Springer Nature Singapore Pte Ltd. 2020
A. Elçi et al. (eds.), *Smart Computing Paradigms: New Progresses and Challenges*,
Advances in Intelligent Systems and Computing 767,
https://doi.org/10.1007/978-981-13-9680-9_3

1   Introduction

Technological evolution has made communications a lot simpler. To take matters into perspective, twenty years ago, people used telegraphic messages and postal letters for communication, which used to have limitations in the amount of data being sent or the exorbitant amount of time it would take (generally, letters took up to a week, sometimes even a month, to reach the receiver). Today, a person sitting in the comfort of his home can send data to the receiver who is at a different part of the world in a mere matter of seconds and what is more, it has become so simple and easy to use that anybody can avail its services.

Wireless networks like wireless local area network (WLAN), long-term evolution (LTE), etc. ensure better connectivity when the nodes involved are mobile. Each access point (AP) provides coverage for a particular area in which the mobile node (MN) is present. This implies that for the MN or the mobile host (MH) to stay connected, there needs to be a handover process that will re-establish the communication channel between the MN and the next best AP which can either be on the same network or on a different network. The method by which connections are renewed and communicational links are re-established between a node/device and an access point whenever the connection gets terminated or the node moves away from the coverage area of the access point is known as handoff/handover.

Switching to APs across different networks is known as vertical handoff while switching to APs on the same network is known as horizontal handoff. Most of the handoff methods proposed by different researchers revolve around this concept. Heterogeneous network selection is currently the most researched field in mobility management techniques as it makes it easier to provide accessibility for nodes in a highly mobile environment as the channels do not have to be on the same network always but can select the most optimal AP, which may be on a different network. The downside to this approach would be that all MNs will need to have the hardware to support these different types of networks.

VANETs are the best examples of networks that support continuous communication mechanisms for mobile nodes in a dynamic environment. VANETs are used in inter-vehicular communications. Two types of communication nodes that can be seen in a VANET environment are vehicle-to-vehicle (V2V) and vehicle-to-infrastructure (V2I) [1]. Handoff is inevitable in such a dynamically changing environment. Of the available handoff methods, re-routing happens to be the widely used handoff method [2]. Section 2 will give an overview about the various handoff categories and the different methods that come under each category, procedures, functions, etc. Section 3 compares the various methods that are discussed in the paper. These sections are then followed by the conclusion. We shall also understand what cognitive radio-based (CR) networks are and how important they are to automatic handoff mechanisms.

2 Overview Various Methods of Handoff

Many handoff methods have been proposed by numerous researchers, which aim at making the process an efficient and effective one. Handoff aims at re-establishing communicational links and reconnecting mobile nodes and access points whenever the mobile node goes out of the coverage area of the access point or cell which is currently providing service. Figure 1 shows a simple handoff process, which will help to understand the method in a simpler manner. When the MN is about to leave the coverage area of the currently servicing access point, it begins to alert the access point which in turn begins to look for suitable target access points (TAP). TAPs are selected based on a number of criteria which may include the distance of the TAP from the MN, the direction in which the MN is travelling, the network selection type, etc.

The method of handoff is categorized into numerous divisions which are based on the number of networks that are involved in the process and the number of access points that the MN is connected to at the same time during handoff, within the network. There are two types of handovers based on the number of networks that are involved in the process. They are horizontal handoff and vertical handoff. Figure 2 represents the categorization of the various methods based on the number of networks involved in the handoff process. Horizontal handoff is when the handover process takes place within the same network while vertical handoff is when the handover process happens across multiple networks.

2.1 Heterogeneous Networks

Heterogeneous wireless networks (HWN) belong to the family of radio-based access networks which are continually evolving [3]. Basically, HWN is made up of numerous wireless networks such as Wi-Fi, WiMAX, LTE, Bluetooth, etc. Every network has its own set of attributes which makes it different from the rest of the available technologies. Attributes like packet rate, delay, latency, energy consumption, range of the network, protocols are part of the network, the amount of space that the network and data relevant to it takes up, etc. are considered. The MN that uses multi-mode devices or CR-based devices will also be able to use the features that accompany those heterogeneous technologies during certain circumstances [4].

Proposals for cells with a lesser area of concentration for effective coverage have been put forth by standards organizations such as 3GPP. These cells include pico-cells, microcells and femtocells and they are effective when it comes to increasing the capacity of the system on the whole, by closing the distance between the APs and the MNs, thus reducing the amount of power required to maintain good communication links. The smaller cells also help the 4G network to migrate towards HWN [5]. The communications structure proposed by IEEE 802.21 standard [6] aids in performing a vertical handoff without losing the channels that were established for

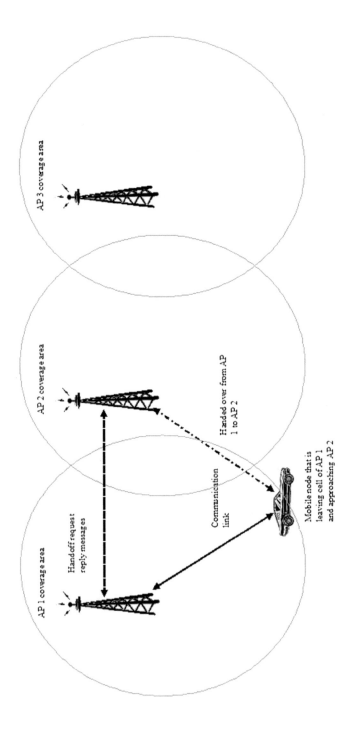

Fig. 1 A simple handoff procedure

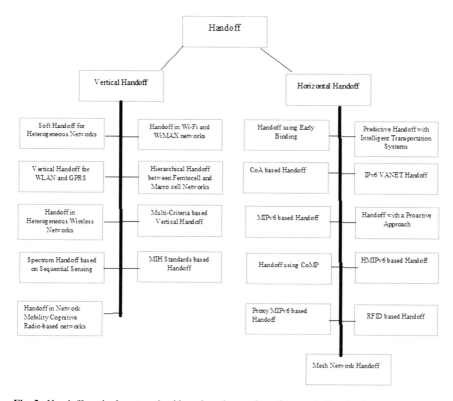

Fig. 2 Handoff methods categorized based on the number of networks involved

communications by simply replacing the type of connection the MN uses to gain access to the infrastructure or AP [3].

2.2 Cognitive Radio-Based Network

CR-based networks, if used efficiently, can lower spectrum utilization [7]. Unlicensed users can also make use of the available spectrum without causing any disruptions for the primary users who are transmitting within their reserved channels, with the aid of CR networks [8]. Features such as spectrum sensing and data transmission functionalities have been added to the physical layer of CR networks [9]. Attributes such as decision-making ability for spectrum, the mobility of spectrum, spectrum sensing and spectrum sharing have been appended to the list of features that the MAC layer of CR networks can boast of [10]. The only difference between the network layer of CR networks and the traditional system is brought about by the fact that the availability of the spectrum affects routing [11, 12].

2.3   Vertical Handoff Methods

The handoff process which takes place between two or more different types of networks is known as a vertical handoff. For example, in a scenario where two networks are available, if an MN needs to connect to an AP for communication, it will decide between the available LTE AP and a Wi-Fi AP by measuring the quality of the connection and the distance between the AP and the MN. This sort of a handover process will be known as a vertical handover as the communication link for the MN toggles between the two available networks.

2.3.1   Soft Handoff for Heterogeneous Networks

Mobile IP addressing schemes and its subset of derived processes are designed to improve the scalability of handoff in IP-based networks using methods such as encapsulation and tunnelling [13]. Functionality of these IP-based networks depends on three basic operations: Care-of-Address (CoA) broadcasting, current location (of the MN) registering and tunnelling [14]. Hierarchical mobile IP (HMIP) is a scheme developed to address the problem faced due to increased mobility which acts as an extension to the base Mobile IP and it enables for extensively scalable and localized architecture. It is controlled with the use of gateway foreign agent (GFA) [14]. Stream control transmission protocol (SCTP), which belongs to the transport layer, is able to operate in wireless IP-based systems (e.g. the Internet). The advantage of SCTP lay in its ability to support multiple hosts from different types of networks. Dynamic host configuration protocol is utilized for assigning IP address, which is known as a CoA, for the network the MN visits [15]. As the MN enters the cell of another access point, away from its own home network, it broadcasts a DHCP '*discover*' message for which the DHCP server replies with a DHCP '*offer*' message. This message contains an IP address which gets assigned to the MN. The MN, in turn, uses this newly assigned IP address as its primary address for newer connections. As the new AP's advertisement is received by the MN, it resends the DHCP '*discover*' message again to which DHCP server responds to with an '*offer*' message. The overhead can be reduced even further if the DHCP server assigns a globally routable IP address to the MN. Upon receiving the '*offer*' message, the MN sends a DHCP '*ack*' reply as acknowledgement for the new IP address assigned. A binding update is then sent to the corresponding node, which uses the new address to reach the MN. The home agent is notified by the MN about its current location. An SCTP extension at the end points enables the reconfiguration of IP addresses such that the MN can request its peers for a new IP address which will enable the network to maintain an active source or destination at all times [14].

2.3.2 Vertical Handoff for WLAN and GPRS

In the method from [16], handoff between WLAN network and GPRS network takes place. To perform a vertical handoff, certain operations such as connection authentication, decision-making and mobility management must be taken care of. Mobility management schemes are of two major categories: routing-based and pure-end system. The routing-based mobility management scheme is used in this method which means that the existing correspondent nodes and applications need not be modified in anyway. Pure-end system approach requires modification in the current layout of TCP/IP as the transport layer and its applications are modified at the end users' part. But this modification may prove to be costly and this is the reason why the focus of this method lies on routing-based scheme. Physical layer attributes such as received signal strength and signal decay are considered as parameters for vertical handoff [17]. Handoff decision can also be made using attributes like packet delay, packet loss, etc. that are considered in this handoff method [16]. Packets travelling in a network make use of the IP addresses provided in order to reach their predestined target. However, allocating IP addresses to MNs will inhibit its mobility because, as the MN is outside of its home network, the assigned IP address would not make sense to the currently servicing AP. Thus, the probability of the IP address being lost will be high. In this method, a seamless handoff proxy server (SHPS) is added to the homer network. Along with this server, a seamless handoff agent (SHA) is installed at the MN to help in the handoff process. A home IP address is assigned to the MN by its home network. The SHA binds the home address of the MN to the virtual NIC used in the network. The MN itself contains two NIC, one for WLAN and the other for GPRS. When the MN moves to WLAN, the SHA binds the IP address the MN receives from the foreign network for the WLAN connection to the WLAN NIC of the MN. This is then registered at the SHPS. Similar operations are performed for GPRS connectivity as well. The MN applications make use of the home address as the source address. The virtual NIC in the SHA accepts the packet and SHA then forwards the packet to SHPS which de-capsulate the packets and sends it to the corresponding node. If there is any node that is ready to be sent to the MN, the corresponding node then forwards it to the home address of the MN [16]. This method does not require any modification of the existing networks.

2.3.3 Handoff in Heterogeneous Wireless Networks

In [18], handoff takes place between networks of various types. The rate of transmission of data can be increased with the use of multi-network vertical handoff as this enables the MN to communicate via the best network, thus improving the overall quality of service (QoS). A vertical handoff (VHO) decision-making method has been proposed in [18] and the main focus of VHO lies in overall load balancing. The VHO method also aims at optimizing the amount of power used. To choose the most optimal network when faced with multiple network options, a path-selection is also included in the method. Controllers which are known as vertical handoff decision controllers (VHDCs) are placed in the APs. The controllers provide the handoff decision function for a set region containing a single AP or multiple APs. IEEE 802.21 or the media independent handover function (MIHF) is used for obtaining the data that form the decision-making inputs.

The link-layer trigger (LLT) becoming active by either the RSS value going below a predetermined threshold or the RSS value of another AP becoming much higher than that of the current AP forms the trigger for the handoff algorithm to begin the process of selecting an optimal network for establishing a communication channel. A slightly modified version of the dynamic source routing (DSR) protocol is used for selecting the path. The selection of the candidate AP is done with the broadcast of route request messages to which the optimal AP replies with a relay message. The routing tables are then updated with this information. In order to prevent the loss of packets and to relaunch route selection in case of a failure in selecting an optimal AP, the modified DSR enables piggybacking of packets, whenever the downlink rate of the channel goes beyond a predetermined threshold [18].

2.3.4 Handoff in Wi-Fi and WiMAX Networks

In [19], a vertical handoff scheme is proposed which enables handoff to take place between two heterogeneous networks, IEEE 802.11 or Wi-Fi which makes use of unlicensed spectrum and IEEE 802.16 or WiMAX which makes use of authorized spectrum [19]. Handoff latency is an issue that occurs during the switching of access points by the MN and two horizontal handoff methods namely hierarchical mobile IP version 6 (HMIPv6) [20] and fast handoff for MIPv6 [21] were proposed as standards by the open standard organization known as the Internet Engineering Task Force (IETF). In the horizontal handoff method, as the MN reaches the edge of the current network's boundary, it sends a handoff preparation request to MAP and MN attaches its location information to this request. The MAP receives the trace information as well as the current location of the MN based on which the closest target access router is selected. After this request, the MAP sends a handoff preparation reply message to the MN with the information about the target access router as well as the Next Link Care-of-Address (NLCoA) which can be used by the MN when it gets connected to the target access router [19]. After the MN receives this information and the target access router, it sends a handoff preparation notification to the MAP to verify if the LCoA provided is valid in the access router. MAP forwards this to the target access router which then replies with an acknowledgement message authorizing the LCoA provided to the MN. After handoff takes place, the data packets are sent to the previous access router as well as the next access router and stored in the latter's buffer. Once the MN is within the range of the target access router, it sends a New Link Attachment (NLA) message to it so that the saved messages can be sent to the MN. A local binding update (LBU) and a local binding acknowledgement (LBAck) [19] messages exchange are performed between the MN and the MAP. In the vertical handoff method proposed, the handoff takes place between Wi-Fi and WiMAX technologies. In this method, velocity of the MN is considered to be the most important characteristic to be used as a parameter for predicting handoff. Handoff is performed when the MN is at the boundary of its current access point and needs to switch the access points. In the event that there happen to be no relevant access point in the vicinity of the MN, the AP of a different network type is selected. As the MN enters the Wi-Fi network's cell, it detects the signal from the access point. A request message for handoff is sent to the access point, along with the MN's details. The AP, if available, will then configure the details according to the information provided by the MN for registering into that network. This configured information is transferred to the MN which then uses the same during handoff which takes place once the MN needs to establish a new link.

2.3.5 Hierarchical Handoff Between Femtocell and Macro-Cell Networks

In [22], a hierarchical handoff that takes place when a MN moves from a macro-cell network to a femtocell network is provided. For a handoff process and the performance of the devices to be optimal, it is necessary to reduce the handoff latency and the number of handovers that take place. Comparing received signal strength (RSS) values of the access points along with the concept of hysteresis and threshold is considered in many handoff algorithms [23]. Hysteresis is said to be a system's property of being dependent on its different states. Whenever a handoff needs to be performed between macro-cells and femtocells, a macro-cell will give first preference to femtocells to improve femtocell utilization as well as for optimal use of different billing models [22]. RSS values from the currently serving macro-cell and target femtocell base stations are combined. The combination process is given by

$$s^\alpha{}_{\text{pro}}[k] = \bar{s}_f[k] + \alpha \bar{s}_m[k] \tag{1}$$

where $\alpha \in \leftarrow [0, 1]$ denotes a combination factor introduced to compensate for the large asymmetry in the transmission powers of both cells.

An offset which proves to be adaptive, which is determined by the RSS values of the macro-cell base station and the combination factor, is the interpreted outcome of the combination process. RSS value from the femtocell base station is compared with that from the macro-cell base station, which has its RSS value scaled by $1 - \alpha$ [22]. A threshold which is given by sf, th is used to see if the RSS value from the femtocell base station is greater than the threshold value. If the RSS value is greater, only then the combination process is applied. This way, the threshold can be controlled strategically in order to select the best possible connection between the available macro-cell base stations or femtocell base stations to ensure better quality of service (QoS) during handoff. Optimal combination factor value plays a vital role in this algorithm. To ensure that the selected combination factor is the optimal value, the following method is used.

$$\alpha^* = \arg\max_{\alpha \in [0,1]} \frac{P_f[k_0+k_1] - P_f[k_0 - k_1]}{2k_1} \tag{2}$$

which is subject to $Pr\{M(k_0) \text{ and } \bar{s}_f[k_0] > s_{f,th}\} < \in$

Where $Pm[k]$ and $Pf[k]$ stand for the probability of the MN to be assigned to either macro-cell base station or femtocell base station at a given time k, respectively. $M(k)$ and $F(k)$ denote the events. k_0 and k_1 stand for indexes that correspond to the femtocell boundary and the marginal distance. These values help in realizing the speed at which the actual handoff process takes place at [22]. This equation means that we are able to choose the combination factor which can be considered to be optimal or to give a value for a small amount of delay, around the boundary of the cells [22].

2.3.6 Multi-criteria-Based Vertical Handoff

The method proposed in [24] considers signal-to-interference and noise ratio (SINR) as the criteria for the predicting the optimal network in a vertical handoff (VHO) process. In forming radio links, different QoS parameters are considered and evaluated. Generally, received signal strength (RSS) forms the crux of VHO methods as seen from methods put forth in [25]. However, as RSS is not a QoS-based schema, it can be used in scenarios that require quality over speed. A SINR-based vertical handoff deals with providing QoS solutions.

$$R_{AP} = W_{AP} \ \log_2 \frac{(1 + Y_{AP})}{T_{AP}} \tag{3}$$

$$R_{BS} = W_{BS} \ \log_2 \frac{(1 + Y_{BS})}{T_{AP}} \tag{4}$$

Shannon's capacity formula states that the maximum achievable data rate from the access points and the base stations are represented by the Eqs. (3) and (4). Where $W_{AP} = 22$ MHz [26] and $W_{BS} = 5$ [27] MHz are carrier bandwidth values [24].

T_{AP} and T_{BS} are the channel coding loss factors.

Based on Eqs. (3) and (4), another equation can be made with the assumption that both the data rates are similar or equal.

$$Y_{AP} = T_{BS} \left(\frac{(1 + Y_{AP})}{T_{AP}} \frac{W_{AP}}{W_{BS}} - 1 \right) \tag{5}$$

This relationship aids in making the SINR-based VHO method an applicable one by having the received SINR value and converting it to the corresponding Y_{AP} value. The multiple metrics that are involved in the vertical handover decision-making module includes several QoS components such as SINR thresholds, outage possibilities, residual capacities, cost, etc. which are utilized based on a scoring system [24].

2.3.7 Spectrum Handoff Based on Sequential Sensing

In [28], a spectrum handoff method helps secondary users to vacate the primary channels and continue transmission on a different channel which is selected based on sequential sensing of the available spectrum. The method also aims at reducing the number of handoff that takes place for every secondary user transmission [28]. Rather than simply performing sequential sensing of all channels, candidate channels are selected for every secondary user such that an idle channel that has no other secondary users vying for it can be found with the least amount of overhead and computational complexity [28]. At first, as the unlicensed secondary user transmits in the default channel, it is assumed that the transmission that is currently happening over the default channel will continue for a given amount of residual idle time of the current channel, which is predetermined. When the idle time duration available for the current channel goes below a given threshold, it means that the primary user will occupy that channel in a short amount of time and the secondary user will have to be ready for spectrum handoff. The secondary user then performs sequential sensing in his own order for identifying an optimal target channel with maximum idle time. Transactions by the secondary user are then continued in the target channel which has been selected. These steps may be repeated if a primary user needs to the channel before the transactions by the secondary user are completed.

2.3.8 MIH Standards-Based Handoff

IEEE 802.21 proposed media-independent handover (MIH) standards. This was aimed at providing the necessary aid for handover between networks of different kinds by taking into consideration signal strength of an AP as its only criteria for TAP selection. In addition to the attributes that are considered by the MIH standards, a few more attributes like the mobility of the node and network environment are taken into consideration by the method in [29]. Data from the application layer and the link layer are used for taking decisions regarding handoff and this dual layer information sharing is to keep the MN connected at all times and to make the transition during handoff as smooth as possible. Handoff initiation [HI] and handoff decision [HD] making form the different modules of the method from [29] where HI cuts down on the number of handovers within a network which in turn helps in improving the overall throughput of the communication channel while also decreasing the number of packets lost. HI also deals with selecting the most optimal TAP. Similar to [18], link-up, link-down, link-going-down events activate the handover algorithm. HD part of the handover method is responsible for selecting the candidate AP from a list of potential TAPs. If there is only one network that is available, then the candidate is selected from that network and the handover algorithm is executed and if there is more than one network, then the most optimal network is selected based on packet loss ratio (PLR) value, average throughput, latency, etc. This method of handover reduces ping-pong effect, which is basically frequent handovers taking place within the network due to failure in selecting an ideal TAP.

2.3.9 Handoff in Network Mobility Cognitive Radio-Based Networks

A CR-based handoff by making use of the concept of network mobility (NEMO) can be seen in [30]. Generally, CR networks are intelligent networks which are capable of identifying channels which are occupied and which are not, such that the primary licensed users and the secondary unlicensed users can both transmit at the same time within the same spectrum without causing any sort of interference to each other. Multiple attributes decision-making (MADM) methods such as cost-based methods and grey relational analysis (GRA) methods are made use of in order to identify the best network for communication at a particular point in time when there are multiple networks to choose from, so as to maintain good quality during data transmission. Cost-based methods take the attribute values directly and therefore, it is easier to use this system for the MADM method. The steps involved in the algorithm are as follows:

1. Different attribute values are selected and converted into a matrix. $D_{ij} =$ (Data Rate, PLR, Jitter, Price, Traffic Density, Direction, Power Consumption), Where $i =$ access network and $j =$ multiple attributes.
2. The different components are then converted into cost attributes (those which need to be minimum) and benefit attributes (those which need to be maximum). The values derived are based on cost-based method, GRA method and entropy method.
3. Normalized attributes are derived from vector normalization for cost-based methods and max-min normalization for GRA method.
4. Weights are calculated using the entropy method.
5. Grey relational analysis methods and cost-based methods are used to identify the optimal target network [30].

2.4 Horizontal Handoff Methods

Handoff that takes place within the same network is known as horizontal handoff. It is a lot simpler than vertical handoff as it does not need to include the complex procedure of selecting the optimal candidate in an optimal network for handoff. Here, a few horizontal handoff methods which were proposed are discussed.

2.4.1 Handoff Using Early Binding

Early binding method for handoff in wireless mobile IPv6 (MIPv6) is proposed in [31]. MIPv6 were introduced in order to increase the efficiency of MNs as they were almost always in a dynamically changing state or a highly mobile state. This allowed the nodes to communicate much faster without having to spend time in sending and receiving request–reply messages that authenticated channels for communication. Duplicated address detection (DAD), CoA, mobility detection and binding updates are the components of MIPv6 [32]. The aim of the method in [31] is to improve the efficiency of handoff in MIPv6 by reducing the total latency that comes about during the handover process and this can be done by reducing or eliminating the latency in DAD and mobility detection. To ensure seamless transition by anticipation method and reduced latency during handoff, the MIPv6 scheme for communication has been slightly modified such that network layer handoff process takes place first and then data link layer handoff will take place. The role of early binding during handoff is to ensure that the MNs receive information about the next or target AP as soon they are within the coverage area of that AP, before sending trigger messages to its network layer for performing handover. Early binding thus supports the MN by providing it with more amount of time for communications during its stay within the boundary of the target cell. This ensures that the slow-moving MNs need not hurry as they have enough time for information exchange. Flags are used to identify if the MNs are slow-moving or fast-moving and the AP can also identify if the nodes are stationary or not by comparing the time taken to cross the boundary of the cell and the amount of time it stayed in the previous cell [31].

2.4.2 Predictive Handoff with Intelligent Transportation Systems

The method in [33] is one that predicts the location of the MN within a campus area network (Wi-Fi) and enables predictive handoff using intelligent transportation systems (ITS). A neighbour-based technique known as the neighbour graph was brought forth by [34], which ensures that all of the stations are equipped for low mobility conditions by dynamically capturing the pre-positioning stations. A GPS-based scheme for IEEE 802.11-based networks was put forth in [35], which stored the coordinates of the neighbouring nodes are stored in a table to aid the MN during handoff decision-making process. Handoff conventionally involves steps such as scanning, authentication, association, etc. [36]. As the MN comes into the coverage area of another access point, its location is updated in the corresponding access router through base station [33]. Two zones are used to segregate the connections based on the base station range along with two threshold points as well. They are zone 1 or $Z1$ which has good range and zone 2 or $Z2$ which has average range. Threshold values have been defined which determine whether the MN must be handed over or can continue with the same access point. When the MN enters the first zone, it has got good coverage from the access point. As it enters zone 2, it begins to experience average connectivity and starts receiving signals from the other base stations. $Z2$ is known as the overlaid area and once the MN enters this zone, a beacon message or signal with a management frame is received by it [33]. Once the node receives the beacon signal, it updates the dynamic channel scanning table of the corresponding BS with the neighbour details such as MAC address, channel number and capacity. The node will then send a roaming notification signal to the access router through the base station, along with the device information and the target base station details [33].

2.4.3   CoA-Based Handoff

The method in [37] makes use of CoA for handoff. The advantage of using CoA is that it allows for the allocation of a duplicate address to the MN before the handoff process and this in turn reduces the overall delay. Internet protocol (IP) ensured that the network gateway (NG) knew that a handoff was going to take place before the handoff could even occur and this enabled the gateway to store necessary information such as the current node, current access point (CAP) and TAP. NEMO is necessary to ensure smooth transitions between the networks and for carrying the sessions created for other nodes forward when the connections switched to static APs [37]. All MNs in NEMO have two addresses. One is the home address which is assigned by the home network agent (HA) and the other is the CoA, which is assigned by the APs to which the nodes get connected to. Whenever there is a switch, the CoA assigned to the node will get updated in its relevant home registry. Mobile anchor points (MAP) control the routers and assign CoA to the MNs. Each mobile router will contain two addresses. The first one is regional care-of-address (RCoA), which happens to be the address of the MAP of that particular domain, and the second type is onlink care-of-address (LCoA) which is the router's address [37]. The MAP will know if a handoff is going to take place if the mobile router exhibits some sort of motion. But the MAP will not know if it will be an inter-domain handoff or an intra-domain handoff. Because of this, the MAP will send out messages that have the RCoA and the LCoA of the router that is mobile to other routers and this is done in order to identify individual nodes when there are multiple nodes involved in the handoff process. Inter-domain address tables are created and maintained by all of the MAPs in vicinity and this table contains the LCoA of all the mobile routers. Intra-domain tables containing columns such as the current LCoA of the mobile router, ID denoting the IP address of access points and the LCoA of the mobile router under every other router are created by the MAP in whose boundary the mobile MR is present. By utilizing limited broadcast routing (LBR), the table is sent across to all of the routers who cross-check if the values in the table matches its IP address. The matching row is then selected and updated into a fast access memory [37]. This incessant handoff method thus reduces the overall load on the system.

2.4.4 IPv6 VANET Handoff

A new IPv6 format for handoff can be seen in the method from [38]. To analyse and predict the approach of a TAP, a technique known as angle of arrival (AoA) is used by the CAP. The AoA of the MN or other access points are calculated by making use of a static infrastructure as the point of contact. To ensure a seamless handover, the network layer first handed off to the TAP while the data link layer is still connected to the CAP. The information at the router which is controlling the CAP and the TAP is modified whenever a handoff occurs. Then the focus is on the mobility handoff of the MN, between the two cells that are being addressed [39]. When the MN reaches the boundary of the TAP, it pings an MN that is currently connected to the TAP about its approach and that MN in turn alerts its router about the oncoming MN. Neighbouring vehicles are used for adapting cost-cutting measures and to reduce complications.

2.4.5 MIPv6-Based Handoff

The method from [40] focuses on a handoff method for MNs using MIPv6. In such a system, every MN, AP and the routers that control them are all assigned with unique global IPv6 addresses by network operators. A pinging packet or message known as the vehicular router solicitation (VRS) message is used like a proxy [39] to alert the CAP if the MN is about to leave its boundary. All of the AP IDs, MN IDs and the router IDs are shared in the form of handover assist information (HAI), which goes along with the proxy message. The optimal TAP is then selected based on the information which is shared by the HAI, leading to efficient handoff. A proxy advertisement message is then sent to the MN containing the IP address of the TAP while the TAP receives a handover initiation message that contains the IP address of the MN to which it replies with a handoff initiation acknowledgement message. Binding update request is sent across in the meantime by the CAP to the home agent of the MN and the other corresponding nodes along with information such the TAP's IP address. TAP begins to buffer the packets meant for the MN that are being forwarded by the CAP through a bidirectional route that is set up. Once the acknowledgement for the binding update request is received, to ensure seamless transition, the MN detaches itself from the CAP's data link layer and gets attached to the TAP. The buffered packets that are stored at TAP are now forwarded to the MN [39].

2.4.6 Handoff with a Proactive Approach

A proactive approach to handoff can be seen in the method from [41]. It makes use of AP graphs within multiple MNs, thus improving the latency. Context and state transfer methods amongst the APs are dependent on the inter-access point protocol

(IAPP). A directed graph $G = (V, E)$ where V stands for the APs and E stands for the edges that connect the APs is defined. Directed graphs are used as MNs are mobile and move along a particular direction. Spanning trees are used to identify the shortest path within the network. Once the topological ordering graph has been derived, depth first search (DFS) is computed in order to make sure that all of the AP components which are connected strongly can be selected as the output. Then propagation control and elimination of APs from candidacy is completed. The final section of the algorithm deals with the insertion of real-world AP positions into the directed graph that is generated [41].

2.4.7   Handoff Using CoMP

A coordinated multiple point (CoMP) transmission within femtocell networks is proposed in the method from [42]. Femtocell networks are used in the current technological scenario in order to skip the problems which ail conventional communication networks. These networks help in seamless improvements over existing networks and their components by providing superior coverage, improved signal-to-interference-plus-noise ratio (SINR) and optimal energy usage. CoMP was part of the 3GPP Release 12 for LTE Advanced [43] and it has since been researched on for usage in high-speed femtocell networks. The method from [42] is an eNodeB-to-eNodeB process which is used for handoff along with the help of CoMP. evolved node B (eNodeB) is an air medium for 4G LTE, which is basically a communication link between two base stations. A moving train is considered to be the fast-moving femtocell node in this scenario. A central control femtocell (CCF) AP will control the MN and the CAP, which is basically a transceiver mounted on top of the train, whenever handoff needs to be performed, thus controlling the information transfer from the MN as well [42]. As the train reaches the boundary of the serving eNodeB, the signal strength wanes away and the transceiver mounted on the outside of the train begins the handoff process by looking for a suitable candidate AP around. The CCF will disconnect the users of the first compartment from the current transceiver and reconnect them to other transceivers which are still connected to the eNodeB. Once the optimal eNodeB is selected, CCF enables the outside transceiver to communicate with the target eNodeB and the other users within the compartment reconnect to this outside transceiver. When the second compartment follows through and reaches the boundary of the servicing AP, it performs handoff and gets connected to the eNodeB to which the first compartment's transceiver is connected to. Thus, there is no need to waste time in searching for an optimal eNodeB.

2.4.8   HMIPv6-Based Handoff

In [44], layer-3 mobility management technique for vehicular wireless networks is proposed. Mobility management in mobile IPv6 is categorized into localized mobility management and global mobility management. But MIPv6 does not cater to the requirement for handling localized mobility in particular and instead makes use of the same mechanism it employs for global mobility to address the issue of handling local mobility [44]. A node known as mobile anchor point (MAP) is used as a local agent to aid in the handoff process within localized mobility. MAPs are used instead of the foreign agents in MIPv6 and they can be contained anywhere in the hierarchy of routers. In stark contrast to the MIPv6 foreign agents, the MAP is not required to reside on each subnet. The parameters based on which the mobile node (MN) selects the MAP is not defined and are left open in HMIPv6 [45]. A proxy mobile IPv6 protocol was made a standard by the IETF. Contrary to MIPv6, PMIPv6 is implemented by the network itself as it notes the movements of the MN and control mobility management [46]. In the method put forth in [44], the existing HMIPv6 is modified in such a way that the MN can calculate the distance between itself and the target candidates which may become its MAP based on the requirements, autonomously. Thus, this method disallows the MN itself to track the MAP node rather the tracking is done through a secondary device.

Due to the scheme being implemented in the network layer, the target nodes can be identified much before the actual signals can be sensed [44]. Here, MAP nodes are replaced even while the existing MAP nodes are valid, thus avoiding loss of packets. The idea here is to make use of a topology learning algorithm along with a path prediction method to enhance the forwarding strategy. The method consists of two basic entities known as the home server (HS) and the forwarding server (FS). Generally, HMIPv6 and PMIPv6 dictate that the mobile node can bind to the next MAP only after it has crossed the boundary of the first MAP or has connected to an AP that belongs to the second MAP and due to this hard handoff method of switching the communication links, the connection is temporarily cut. In this method, a soft handoff is performed where the MN is able to understand the topology of the network, by making use of the topology learning algorithm, and enable the handover process according to the available MAP nodes [44]. MN is not required to be aware of the entire network topology and it can instead learn about the topology from the different APs as the MN moves around. The network topology learning algorithm is made up of definitions and theorems which are based on the number of hops, the connections that exist between the nodes, the distance between the nodes, etc. When the distance between the forwarding server and the MN exceeds a pre-defined threshold, a new FS is selected by the MN, which happens to be the closest FS.

2.4.9   Proxy MIPv6-Based Handoff

In [47], we see a method that uses proxy MIPv6 technique to perform handoff. Packet loss ratio (PLR) increases due to a large loss of data packets when frequent handovers occur in a mobile environment. This will also increase latency in the network. As seen in previous methods, making use of early binding updates can reduce latency and delay in handing over, thus preventing the loss of packets. A mobile access gateway (MAG) is responsible for maintaining a collection of IP addresses which are assigned by a network administrator. The gateway also maintains a table with information about other gateways or routers within the network and this is helpful because, when a new TAP needs to be found for handoff, the table can be looked up by the router and easily identify an eligible TAP. Once the target gateway has been found, the source gateway will notify the target gateway about the approaching MN such that the next gateway can get ready for handoff by assigning a local IP address for the MN. At the local mobility anchor, a binding update entry will be made [39]. An information request acknowledgement (IRA) will be sent by the gateway to the MN which is coming into its coverage area so that the MN can sync its IP address with that of the target gateway while still connected with the source gateway. To help the method in identifying the direction in which the MN is travelling, GPS is used to share information about the location of the MN [47].

2.4.10 RFID-Based Handoff

The method from [48] uses RFID tags in order to perform the handoff operation. This method was proposed because it was identified that vehicles suffered zero breakage in communication channels given that the APs and the MNs were static. Due to the highly mobile environment of vehicular networks, it is necessary for a system that performs fast handoff. Each vehicle's chassis is fitted with an RFID tag and is assigned a unique identification number called the MAC address. Scanners for RFID tags are kept along the roadside and these scanners identify the vehicles that pass by. The scanned MAC addresses are then forwarded to the access gateways and they are responsible for performing handoff operations. The access gateway is also responsible for controlling a group of APs within a subnet and therefore, whenever a MAC address is identified, the gateway will know which vehicle is approaching the subnet. This reduces the overall delay in performing handoff. Local vehicle server (LVS) is used by the subnet in order to maintain information about all of the MN so that the efficiency can be improved [48]. LVS contains information such as the MAC address of the vehicles, the IP addresses of the vehicles and the AP IP addresses. Yet another server known as the global vehicle server (GVS) stores information about the location of the vehicles, the IP addresses of the vehicles and the IP addresses of the access gateways. The home address of all MNs are assigned by the HA. GVS also maintains a separate table known as the vehicle address relation table which stores the MAC addresses and the home addresses of the MNs.

2.4.11 Mesh Network Handoff

In [49], we see a method which aims that performing handoff within a wireless mesh network. A home mesh router (HMR) is basically the coverage area of a cell within which a node resides. Whenever the node leaves the HMR, the handoff process is started. The selection of a candidate router or the foreign mesh router (FMR), which is basically any router that is not its home router, is based on the SINR values of that router. To get connected to any candidate FMR and to begin communication, the node needs to be authenticated by the FMR in question. Due to the time taken for the FMR to authenticate the MN, latency and overall delay increase. To avoid delays, [49] suggests the use of handoff authentication techniques such as public key-based and symmetric key-based methods which issue tickets, in order to reduce delay during authentication period. Between a mesh client and a mesh server, a master key is generated and between selected mesh routers, a group master key is generated to aid in faster authentication processes. All routers receive tickets generated by servers for respective mesh routers. Whenever the roaming client or the MN is able to match the key it has with the publicly available key, handoff will be initiated.

3   Analysis and Comparison of Methods

Every method discussed in this study follows a different approach for performing handoff operations. Some of the methods try to reduce the overall latency, some try to reduce the amount of power consumed, some methods try to improve the speed with which connections are established, while some others try to establish a method for selecting optimal networks from a list of available networks. The different methods and the various attributes they focus on can be seen in Table 1. We can also see the approach each method follows.

4   Conclusion

The study's goal is to compare various handoff methods that were proposed by numerous researchers. Every method discussed in this paper has its own set of key attributes and unique approaches to optimize the handoff process. Identifying the key features of each process is what the paper aims at. The different methods have been categorized based on the number of networks involved, viz. vertical handoff and horizontal handoff. It has been identified that vertical handoff methods are prevailing in providing seamless connectivity across different networks. Commercially viable applications are being designed based on the numerous vertical handoff strategies available. Every method has its advantages and disadvantages which can be observed

Table 1 Comparison of various methods of handoff discussed in the paper

Name	Approach	Horizontal handoff	Vertical handoff	Seamless	Latency	Overhead optimization	Delay reduction	Location based	Packet loss
Vertical Handoff for WLAN and GPRS	Routing-Based Approach with a Virtual NIC instead of a foreign agent		Yes	Yes					
Soft handoff for heterogeneous networks	A protocol that operates over IP stream control transmission		Yes	Yes					
Predictive handoff with ITS	Makes use of signal strength, threshold based on the signal strength and its capacity, along with the location of the mobile node for predictive handoff within a campus area Wi-Fi network	Yes		Yes	Yes			Yes	

(continued)

Table 1 (continued)

Name	Approach	Horizontal handoff	Vertical handoff	Seamless	Latency	Overhead optimization	Delay reduction	Location based	Packet loss
Handoff in Wi-Fi and WiMAX networks	A horizontal method as well as a vertical handoff method that is based on the mobile node's location and its movement pattern	Yes	Yes		Yes				Yes
Hierarchical handoff between femtocell and macro-cell networks	A hierarchical approach where the RSS values from the available macro-cell base stations and femtocell base stations are combined to generate an adaptive offset that can be used to select the optimal base station		Yes					Yes	

(continued)

Table 1 (continued)

Name	Approach	Horizontal handoff	Vertical handoff	Seamless	Latency	Overhead optimization	Delay reduction	Location based	Packet loss
Multi-criteria-based vertical handoff	It used SINR as the criteria for predicting the optimal network in a vertical handoff process. different QoS parameters are considered in forming radio links		Yes	Yes				Yes	
HMIPv6-based handoff	A topology learning algorithm in HMIPv6 that uses MAP replacements to perform soft handoff in the network layer	Yes		Yes			Yes	Yes	Yes

(continued)

Table 1 (continued)

Name	Approach	Horizontal handoff	Vertical handoff	Seamless	Latency	Overhead optimization	Delay reduction	Location based	Packet loss
Spectrum handoff based on sequential sensing	Sequential sensing spectrum handoff for multiple users which helps in spectrum handoff when there are multiple secondary and primary users occupying same channels		Yes	Yes		Yes			
Handoff using early binding	Anticipatory method by making use of early binding	Yes		Yes	Yes				

(continued)

Table 1 (continued)

Name	Approach	Horizontal handoff	Vertical handoff	Seamless	Latency	Overhead optimization	Delay reduction	Location based	Packet loss
Handoff in heterogeneous wireless networks	Vertical handoff for selecting optimal networks amongst various types by implementing vertical handoff decision-making algorithm across various vertical handoff decision-making controllers		Yes			Yes			

(continued)

Table 1 (continued)

Name	Approach	Horizontal handoff	Vertical handoff	Seamless	Latency	Overhead optimization	Delay reduction	Location based	Packet loss
CoA-based handoff	Incessant handoff. Makes use of pre-allocation method to allocate the address to each mobile node. Care-of address is used to identify the mobile node whenever it requests for handoff to a particular access point	Yes		Yes			Yes		Yes
IPv6 VANET handoff	Angle of arrival is used by a static infrastructure to identify and predict approaching vehicles	Yes		Yes					

(continued)

Table 1 (continued)

Name	Approach	Horizontal handoff	Vertical handoff	Seamless	Latency	Overhead optimization	Delay reduction	Location based	Packet loss
MIPv6-based handoff	Vehicular route solicitation is shared as a proxy message so that it announces its readiness for handover to the access points and it contains handover assist information which contains the various addresses that help during handoff	Yes		Yes					Yes

(continued)

Table 1 (continued)

Name	Approach	Horizontal handoff	Vertical handoff	Seamless	Latency	Overhead optimization	Delay reduction	Location based	Packet loss
Handoff with proactive approach	The method makes use of access point graph for a proactive approach that scans only the selected AP based on association patterns	Yes		Yes	Yes				
Handoff using CoMP	eNodeB-to-eNodeB handoff method that makes use of coordinated multiple point transmission for femtocell networks	Yes		Yes	Yes	Yes	Yes		Yes

(continued)

Table 1 (continued)

Name	Approach	Horizontal handoff	Vertical handoff	Seamless	Latency	Overhead optimization	Delay reduction	Location based	Packet loss
Proxy MIPv6-based handoff	The mobile access gateway maintains a pool of IP addresses of all the nodes. Previous mobile access gateway identifies the next mobile access gateway from the table of gateways available. Also makes use of GPS to identify the node's path	Yes		Yes	Yes		Yes	Yes	

(continued)

Table 1 (continued)

Name	Approach	Horizontal handoff	Vertical handoff	Seamless	Latency	Overhead optimization	Delay reduction	Location based	Packet loss
RFID-based handoff	RFID tags are used in this method which are embedded into the vehicular nodes and these tags contain MAC addresses of the vehicles. RFID scanner strips along the roadsides are used to scan these RFID tags and these scanners further forward the MAC addresses to the access gateways for faster handoff process	Yes		Yes	Yes		Yes	Yes	

(continued)

Table 1 (continued)

Name	Approach	Horizontal handoff	Vertical handoff	Seamless	Latency	Overhead optimization	Delay reduction	Location based	Packet loss
MIH standards-based handoff	Media-independent handover is considered in this method with added parameters		Yes	Yes			Yes		Yes
Mesh network handoff	Keys are used predomi-nantly in this system where public keys and symmetric keys can improve the efficiency of handoff. Group master keys are shared amongst a group of selected mesh networks	Yes			Yes	Yes	Yes		

(continued)

Table 1 (continued)

Name	Approach	Horizontal handoff	Vertical handoff	Seamless	Latency	Overhead optimization	Delay reduction	Location based	Packet loss
Handoff in network mobility cognitive radio-based networks	Multiple attributes decision-making methods like cost-based method and grey relational analysis methods are employed here which considers various attributes of networks for optimal selection		Yes	Yes	Yes		Yes		

from the above discussions and results. It can be noted that numerous methods focus on optimizing a single attribute of the networks involved. Most the methods make use of only a single attribute such as signal strength, in order to perform handoff when in reality, numerous attributes play vital roles in optimizing the handoff process. In the future proposing a handoff method by identifying winning features from a select few methods needs to be done. Multi Attributes Decision Making (MADM) methods can be considered in order to improve the overall efficiency of a handoff process.

Acknowledgements The authors would like to thank the Dept. of Computer Science at Christ University, Bengaluru, India for their whole- hearted support.

References

1. R. Sharma, J. Malhotra, in *A Survey on Mobility Management Techniques in Vehicular Ad-hoc Network*. International Conference on Computing Communication and Systems, 2014, pp. 38–41
2. X. Wang, D. Le, Y. Yao, A Cross-layer mobility handover scheme for IPv6-based vehicular networks. AEU-Int. J. Electr. Commun. **69**(10), 1514–1524 (2015)
3. S.M. Matinkhah, S. Khorsandi, S. Yarahmadian, A load balancing system for autonomous connection management in heterogeneous wireless networks. Comput. Commun. (2016)
4. E. Avelar, L. Marques, D. Passos, R. Macedo, K. Dias, M. Nogueira, Interoperability issues on heterogeneous wireless communication for smart cities. Comput. Commun. **58**, 4–15 (2015)
5. Y. Li, B. Cao, C. Wang, Handover schemes in heterogeneous LTE networks: challenges and opportunities. IEEE Wirel. Commun. **23**(2), 112–117 (2016)
6. K. Taniuchi, Y. Ohba, V. Fajardo, S. Das, M. Tauil, Y.H. Cheng, A. Dutta, D. Baker, M. Yajnik, D. Famolari, IEEE 802.21: media independent handover: features, applicability, and realization. IEEE Commun. Mag. **47**(1), 112–120 (2009)
7. FCC, Promoting more efficient use of spectrum through dynamic spectrum use technologies, Federal Communication Commission (FCC 10-198), Washington DC, Tech. Rep. 10-237, Nov 2010
8. N. Kaabouch, W.-C. Hu (eds.), *Handbook Of Research On Software-Defined and Cognitive Radio Technologies For Dynamic Spectrum Management*. (IGI Global, 2015)
9. D. Cabri, S. Mishra, D. Willkomm, *A Cognitive Radio Approach For Usage Of Virtual Unlicensed Spectrum*. 14th IST Mob., 2005
10. A. Sultana, X. Fernando, L. Zhao, An overview of medium access control strategies for opportunistic spectrum access in cognitive radio networks. Peer–to-Peer Netw. **10**(5), 1113–1141 (2016)
11. I. Akyildiz, W. Lee, K. Chowdhury, CRAHNs: cognitive radio Ad hoc networks. Ad Hoc Netw. **7**(5), 810–836 (2009)
12. M. Bouabdellah, N. Kaabouch, F.E. Bouanani, H. Ben-Azza, Network layer attacks and countermeasures in cognitive radio networks: a survey. J. Inf. Sec. Appl. **38**, 40–49 (2018)
13. C. Perkins, *IP Mobility Support for IPv4*, RFC 3220, Jan 2002
14. A. Argyriou, V. Madisetti, A soft-handoff transport protocol for media flows in heterogeneous mobile networks. Comput. Netw. **50**, 1860–1871 (2006)
15. R. Droms, *Dynamic Host Configuration Protocol*, RFC 2131, Mar 1997
16. R. Jan, W. Chiu, An approach for seamless handoff among mobile WLAN/GPRS integrated networks. Comput. Commun. **29**, 32–41 (2005)
17. Q. Zhang, C. Guo, Z. Guo, W. Zhu, Efficient mobility management for vertical handoff between WWAN and WLAN. IEEE Commun. Mag. **41**, 102–108 (2003)

18. S. Lee, Vertical handoff decision algorithms for providing optimized performance in heterogeneous wireless networks. IEEE Trans. Veh. Technol. **58**(2), 865–881 (2009)
19. F. Shi, K. Li, Y. Shen, Seamless handoff scheme in Wi-Fi and WiMAX heterogeneous networks. Futur. Gener. Comput. Syst. **26**, 1403–1408 (2010)
20. H. Soliman, C. Castelluccia, K. El-Malki, L. Bellier, *Hierarchical Mobile IPv6 Mobility Management (HMIPv6)*. IETF RFC, 4140, 2005
21. G. Koodli, *Fast Handovers For Mobile IPv6*. IETF RFC, 4068, 2005
22. J. Moon, D. Cho, in *Novel Handoff Decision Algorithm in Hierarchical Macro/Femto-Cell Networks*. WCNC 2010 proceedings, IEEE Communications Society
23. G. Pollini, Trends in Handover Design. IEEE Commun. Mag. **34**(3), 82–90 (1996)
24. G.A.F.M. Khalaf, H.Z. Badr, A comprehensive approach to vertical handoff in heterogeneous wireless networks. J. King Saud Univ. Comput. Inf. Sci. **25**, 197–205 (2013)
25. A. Benmimoune, M. Kadoch, in *Vertical Handoff Between UMTS and WLAN*. Proceedings of the Fourth International Conference On Communications and Information Technology (CIT'10), Corfu Island, Greece, (2010), pp. 131–140
26. P.P. Krishnamurthy, A. Hatami, M. Ylianttila, J. Makela, R. Pichna, J. Vallstom, Handoff in hybrid mobile data networks. IEEE Pers. Commun. (2006), pp. 34–46
27. A.H. Zahram, B. Liang, A. Saleh, Signal threshold adaptation for vertical handoff on heterogeneous wireless networks. ACM/Springer Mobile Netw. Appl. (MONET) J. **11**(4), 625–640 (2006)
28. W. Zhang, C.K. Yeo, Sequential sensing based spectrum handoff in cognitive radio networks with multiple users. Comput. Netw. **58**, 87–98 (2014)
29. U. Kumaran, M.K. Jeyakumar, An optimal vertical handover strategy for vehicular network based on IEEE 802.21 MIH standards. Int. J. Innovations Eng. Technol. **7**(3), (2016)
30. K. Kumar, A. Prakash, R. Tripathi, A spectrum handoff scheme for optimal network selection in NEMO based cognitive radio vehicular networks. Wireless Commun. Mobile Comput. (2017), Hindawi Publishing Corporation
31. H. Kim, Y. Kim, in *An Early Binding Fast Handover for High-Speed Mobile Nodes on MIPv6 over Connectionless Packet Radio Link*. Proceedings of 7th ACIS International Conference on Software Engineering, Artificial Intelligence, Networking, and Parallel/Distributed Computing (2007), pp. 237–242
32. H. Soliman, *Mobile IPv6: Mobility in a Wireless Internet* (2004), Addison Wesle
33. G. Kousalya, P. Narayanasamy, J. Hyuk Park, T. Kim, Predictive handoff mechanism with real-time mobility tracking in a campus wide wireless network considering ITS. Comput. Commun. **31**, 2781–2789 (2008)
34. A. Mishra, M. Shin, W. Arbaugh, in *Context Caching Using Neighbor Graphs For Fast Handoffs In A Wireless Network*. Proceedings of Twenty Third Conference of the IEEE Communications Society (INFOCOM) (Hong Kong, Mar 2004)
35. G. Singh, R.B. Singh, A.P. Singh, B.S. Sohi, in *Optimised GPs Assisted Low Latency Handoff Schemes For 802.11 Based Wireless Networks*. Proceedings of 13th International Conference on Advanced Computing and Communications (ADCOM 2005)
36. B. O'Hara, A. Petrick, IEEE 802.11 Handbook—A Designer's Companion, 2nd edn. (Mar 2005)
37. M. Dinakaran, P. Balasubramanie, An efficient hand-off mechanism for vehicular networks. J. Comput. Sci. **8**(1), 163–169 (2012)
38. X. Wang, H. Qian, A mobility handover scheme for Ipv6-based vehicular Ad hoc networks. Wirel. Pers. Commun. **70**(4), 1841–1857 (2013)
39. P. Roy, S. Midya, K. Majumder, in *Handoff Schemes in Vehicular Ad-Hoc Network: A Comparative Study*. Proceedings of Intelligent Systems Technologies and Applications 2016, Advances in Intelligent Systems and Computing, vol. 530, (2016), p. 421
40. L. Banda, M. Mzyece, G. Noel, in *Fast Handover Management In IP Based Vehicular Networks*. IEEE International Conference on Industrial Technology (ICIT), (2013), pp. 1279–1284
41. D. Das, R., in *Misra, Proactive Approach based Fast Handoff Algorithm for VANETs*. Proceedings of India Conference Annual IEEE (2013)

42. S. Chae, T. Nguyen, Y.M. Jang, Seamless QoS-enabled handover scheme using CoMP in fast moving vehicular networks. Int. J. Distributed Sens. Netw. (2013), Hindawi Publishing Corporation
43. GPP TR 36.836, *Mobile Relay for Evolved Universal Terrestrial Radio Access (E-UTRA)*, 2012
44. Z. Naor, Fast and reliable handoff for vehicular networks. Ad Hoc Netw. **11**, 2136–2145 (2013)
45. H. Soliman, C. Castelluccia, K.E. Malki, L. Bellier, *Hierarchical Mobile IPv6 Mobility Management (HMIPv6)*, Aug (2005), IETF RFC 4140
46. S. Gundavelli, K. Leung, V. Devarapalli, K. Chowdhury, B. Patil, *Proxy Mobile IPv6* (Aug 2008), RFC 5213
47. A. Moravejosharieh, A Proxy MIPv6 handover scheme for vehicular Ad-hoc networks. Wirel. Pers. Commun. **75**(1), 609–626 (2014)
48. S. Midya, in *An Efficient Handoff Using RFID Tags*. Proceedings of International Conference on Intelligent Communication, Control and Devices, Advances in Intelligent Systems and Computing, vol. 479 (2016), p. 779
49. G. Rathee, H.H. Saini, in *A Fast Handoff Technique in Wireless Mesh Network (FHT for WMN)*. Procedia Computer Science 79. Elsevier (2016), pp. 722–728

OpenFlow-Based Multi-controller Model for Fault-Tolerant and Reliable Control Plane

Nabajyoti Medhi and Dilip Kumar Saikia

Abstract Software-defined networking (SDN), which decouples the control and data planes of the network, is one of the most promising paradigms in the recent networking architectures. The SDN concept relies on a centralized controller. However, a centralized architecture faces the challenges of scalability, availability and fault-tolerance. OpenFlow (OF), the most commonly used southbound interface for SDN, makes the provision of using multiple controllers. In this paper, we propose FOIL, a new Fault-tolerant OpenFlow multi-controller model with ICMP-based lightweight inter-controller communication for achieving fault-tolerance and reliability in the SDN control plane.

Keywords FOIL · Fault-tolerance · ICMP · Lightweight communication · Multi-controller · OpenFlow · Reliable control plane · SDN

1 Introduction

The software-defined network (SDN) has many advantages over the traditional network architecture in terms of flexibility, openness, programmability and cost-effectiveness. SDN has a centralized control architecture which makes the network programmable from a controller device. Single controller is vulnerable to single point failure and puts questions on the reliability of the network. In order to ensure reliability, researchers have been working towards developing a suitable architecture for multi-controller SDN.

OpenFlow (OF) [1] is an interface for SDN to program and control the switches of the network. Using OpenFlow, the controller can read switch statistics and write, delete or modify rules installed in the switches either proactively or reactively. Open-Flow controller exchanges OpenFlow messages with OpenFlow switches over a

N. Medhi (✉) · D. K. Saikia
Department of Computer Science & Engineering, Tezpur University, Tezpur 784028, India
e-mail: nmedhi@tezu.ernet.in

D. K. Saikia
e-mail: dks@tezu.ernet.in

© Springer Nature Singapore Pte Ltd. 2020
A. Elçi et al. (eds.), *Smart Computing Paradigms: New Progresses and Challenges*,
Advances in Intelligent Systems and Computing 767,
https://doi.org/10.1007/978-981-13-9680-9_4

secure channel. OpenFlow [2] provides a basic framework for multiple controllers but keeps the options for detailed architecture open. In this paper, we propose a Fault-tolerant OpenFlow multi-controller model with ICMP-based lightweight inter-controller communication named FOIL for ensuring the reliability of the SDN control plane for DCNs. FOIL provides a mechanism for high reliability and fault-tolerance.

The proposed controller is implemented on a Mininet [3]-based emulated test bed for performance study.

The rest of the paper is organized as follows: Section 2 discusses the OF multi-controller framework and reviews the existing works. Section 3 presents our proposed controller. Section 4 presents the simulation-based performance study of the proposed controller. Section 5 concludes the paper.

2 Background and Related Works

Static configuration between controller and switch creates difficulty in terms of han-dling unpredictable traffic in the network. In such a case, it is required to migrate the switches from a heavily loaded controller to a less loaded one using distributed controllers.

OpenFlow provides a multi-controller [2] model which allows multiple OpenFlow controllers to exist in the same network. In the multi-controller scenario, the roles of the OpenFlow controllers are formulated in three categories, namely—MASTER, EQUAL and SLAVE. A MASTER or EQUAL controller role has complete access, including reading asynchronous messages from the switches and writing or modify-ing actions on the switches except that there can be only one MASTER for a switch, the controller in charge. The EQUAL role is the default role of any controller. A SLAVE controller has read-only access to the switches. It does not receive asyn-chronous messages from the switches and cannot write new rules or modify any rule on the switches. When a new MASTER controller is selected for any switch, it changes the role of the earlier MASTER to SLAVE. As a switch can connect to multiple controllers at a time, the OpenFlow multi-controller model provides the scope to incorporate failure avoidance and controller load balancing. Here, the con-troller handover can occur in a distributed manner but no definite basis and scheme for the controller distribution, load balancing, failure recovery, etc., have been spec-ified. It is therefore important to develop suitable mechanisms for the OpenFlow multi-controller framework to achieve desired reliability and fault-tolerance.

The distributed multi-controller implementation must satisfy two important prop-erties at any given time—safety and liveness [4]. Safety requires each OpenFlow switch to listen to only one MASTER or EQUAL controller for the processing of asynchronous messages and liveness requires that at least one controller is active, either MASTER or EQUAL, for each one of the OpenFlow switches.

Physically distributed multiple SDN controller designs found in the literature can be categorized into two types—logically centralized and logically distributed [5]. In a logically centralized design, there are multiple controllers but each controller has the same global view of the network and works with similar responsibilities. In logically distributed design, controllers are physically as well as logically distributed, and the individual controllers may not have a global view of the network. Instead, each may be responsible for a domain of the network and perform a given set of functionalities.

ONIX [6] is one of the logically centralized distributed OpenFlow controller architecture which provides a distributed control plane with cluster of servers running several ONIX instances. It maintains a network information base (NIB) which contains the entire network state that is shared among all the ONIX instances. The ONIX instances require updating the NIBs through database transactions each time any state information is to be communicated between any two controllers and this makes the overall system slower due to frequent database transactions.

Elasticon [7] is another logically centralized architecture with a cluster of controllers sharing the workload by using a flexible switch assignment and migration scheme with role switching. It maintains the global network view using distributed data store modules. Elasticon shows that load balancing improves the controller performance. However, the issue of controller assignment to the switches of a large network has not been addressed in Elasticon.

Kandoo [8] is a logically distributed architecture with two layers of controllers, namely, the local controllers and a root controller running different applications. The local controller queries switches to detect elephant flows and once detected it notifies the root controller to install or update flow entries on network switches. For inter-controller communication, local controllers must signal the root controller which makes the local controllers dependent on the root controller. A failure in the root controller may lead to failure of the network. Reported results show performance improvements in Kandoo as compared to the traditional OpenFlow.

Logically centralized architectures like ONIX and Elasticon work by storing the global network state in distributed data stores. The latency in accessing the distributed data will, however, grow when the global network state information base becomes large. It is therefore required to minimize the volume of data to be accessed from distributed data stores and inter-controller communication should be made light with minimum use of transactional databases.

In a multi-controller environment, dynamic assignment of the switches to the controllers is conceptually simple but hard to achieve without disrupting normal network operations. Further, frequent migration of a switch among controllers may affect the liveness of controller operations. The existing distributed architectures generally tend to use the popular distributed routing protocols like BGP, OSPF, IS-IS, etc., [9] for inter-controller communications. But, these protocols generate routing overhead due to the broadcast of link state databases. Hence, there is need for a distributed architecture with suitable dynamic controller allocation mechanism for load balancing and failure recovery while making inter-controller communications light.

3   The Proposed OpenFlow-Based Multi-controller Model

Use of multiple OpenFlow controllers in a DCN can enhance the reliability and scalability of the DCN. FOIL provides redundancy in the controller with a cluster of controllers to achieve reliability and distribution of the load among controllers in the cluster to achieve scalability. Lightweight ICMP-based inter-controller communication is used to minimize the control traffic.

3.1   The System Model

We illustrate the system model of FOIL in Fig. 1a.

FOIL requires the maintenance of records of switch IDs in terms of switch datapath IDs (DPID) which are recognized by an OpenFlow controller. A DPID is a 64-bit switch identifier to physically represent a switch in OpenFlow network. The switches in the network are controlled by a cluster of controllers. The cluster normally consists of one MASTER controller and at least one EQUAL and one SLAVE controller. Unlike in high availability clusters HA [10], the cluster in FOIL does not only perform fail-over but also share load among the controllers adaptively.

In FOIL, the whole network is operated by a single cluster of controllers. In order to share global information of the network, network states of the cluster may be shared via distributed database (DDB) instances accessible by all the controllers in that cluster. Each controller in the cluster will access the common keyspace from the shared distributed database (DDB) instances. A DDB such as Apache Cassandra [11] may be used to create distributed database instances. Such a keyspace can be used to store the topology, subnet and controller information, etc., of the cluster. Only MASTER controller keeps the authority of modification access to the keyspaces. The controller cluster maintains two lists: (a) native-controller list: A list of IPs of the controllers currently acting as the MASTER, EQUAL(s) and SLAVE(s) in the cluster. The list is updated for any change that occurs and is made accessible to all the controllers in the cluster through a database keyspace; (b) native-switches list: A list of DPIDs of the switches that are mapped to their assigned MASTER controllers. The list contains the DPIDs of all the switches in the network and their respective MASTERs. The MASTER IPs in the native-switches list are updated whenever there is a change of MASTER in a cluster.

3.2   Working of FOIL Clusters

A FOIL cluster provides for failure recovery in case of any failure of a MASTER controller. It also supports load balancing when there is overload on a MASTER due to increase in flow traffic. Failure recovery is achieved in a cluster with an EQUAL

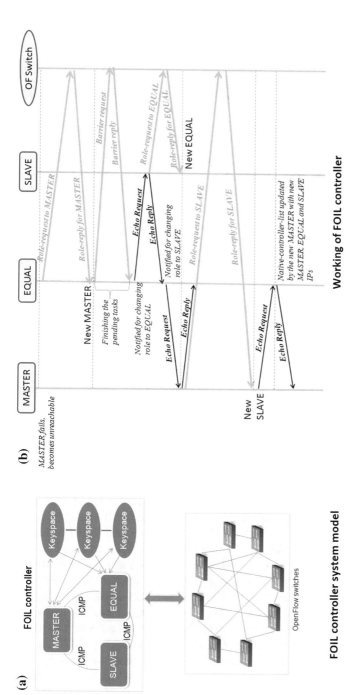

Fig. 1 **a** Illustration of FOIL controller. FOIL controller cluster is connected to all the switches; controllers communicate to each other via ICMP; MASTER can read and write; whereas, EQUAL and SLAVE can only read the DDB keyspaces; **b** switch migration for recovery from the MASTER failure

taking over as the new MASTER. Load balancing on the other hand requires the election of a new MASTER controller before control traffic load exceeds certain CPU and memory thresholds of the MASTER controller. Both failure recovery and load balancing follow the same mechanism but the difference between the two is that—the first one takes care of any unexpected MASTER controller failure that may occur and the later one prevents any kind of failure condition due to MASTER controller overload.

In order to check that the MASTER is active, the EQUAL controllers periodically send echo request messages to it. If an EQUAL controller does not receive echo reply within suitably chosen but randomized timeout period, it again checks for the controller IP in the *native-controller-list* to confirm that MASTER has not changed during its echo request. If there is no change in the MASTER, then it assumes that the MASTER has become inactive. The following steps are taken thereafter:

1. Once the EQUAL realizes that the MASTER is either not active or not reachable, it sends echo request to all the other controllers to notify indicating its intention of replacing the current MASTER, the notification will be sent with a time-stamp. If there was notification received from any other EQUAL, it withdraws from contention.
2. If the current MASTER inactivity was due to some temporary reason, it will respond with the broadcast of an echo request with an ACTIVE message to notify all the controllers in the cluster of its being active.
3. If no ACTIVE message is received within a set period, the other EQUAL/SLAVE controllers verify the validity of the IP address of the requesting EQUAL controller in the native-controller-list. If it is found to be a valid EQUAL controller, the other controllers send echo reply in acknowledgement.
4. If a contender EQUAL receives a notification for replacing the current MASTER from any other EQUAL, it compares the time-stamp in the notification with its own time-stamp. If its own time-stamp is older than the other contender, then it becomes the first contender. Else, it withdraws from the contention.
5. The first contender EQUAL controller on receiving acknowledgements from all other controllers in the cluster, other than the inactive MASTER, starts the following switch migration process as shown in Fig. 1b. The contender EQUAL node sends *role request to MASTER* message to all the OF switches under the control of the inactive MASTER. On receiving a role reply message from a switch, it elects itself as the new MASTER for the switch and issues barrier request message to the switch to finish any pending task at the switch. On completion of the migration of all the switches, the new MASTER notifies the active SLAVE nodes for selection of a new EQUAL by sending an echo request to it. The SLAVE sending the earliest echo reply is selected to be the next EQUAL. The new MASTER notifies the selected SLAVE to change its role EQUAL (not shown in Fig. 1b). The SLAVE immediately on receiving the second echo request sends *role request to EQUAL* message to all the switches. On receiving echo reply from a switch, it becomes an EQUAL node for that switch. The new MASTER now

updates the IP addresses of the new MASTER, EQUAL and SLAVE controllers in the *native-controller-list* of the cluster in the database.

Normally, an EQUAL controller will not respond to any asynchronous message received from the switches in order to satisfy the safety property requirement. But after a MASTER controller failure is detected by any EQUAL controller, it starts processing the asynchronous messages. Multiple EQUAL nodes may process the asynchronous messages till a new MASTER is selected. EQUAL controllers stop processing the asynchronous messages once the MASTER controller IP is changed or updated in the *native-controllers-list* of the cluster. This is to avoid loss of asynchronous messages without being processed during a MASTER controller failure.

Similarly, a SLAVE controller checks whether at least one EQUAL is active. It sends echo requests to all the EQUAL controllers and if it does not receive reply from any of them within the set timeout period, it sends echo request to all other SLAVE controllers indicating its intention of becoming an EQUAL controller. After receiving an acknowledgement from all the SLAVES, the requesting SLAVE sends role request to EQUAL to all the switches. In order to prevent any other SLAVE in that cluster from switching its role to EQUAL, the newly elected EQUAL stops responding to echo request messages from another SLAVE.

Thus, FOIL takes a distributed approach to failure detection and recovery. In this process, the hierarchy of the controller roles is maintained in the cluster. It is ensured that an EQUAL controller can only become a MASTER. A SLAVE has to become an EQUAL first before becoming a MASTER.

4 Performance Evaluation

The performance study of FOIL was carried out on an emulated set-up on an IBM Server x3500 M4 model having Intel(R) Xeon(R) CPU E5-2620 2.00 GHz processor. A three-level tree-based topology was created for experiments in Mininet comprising one core switch, two aggregation switches, n number of access switches connected to each aggregation switch and a single host connected to each access switch.

An experimental test bed is created using Mininet with OpenVSwitch (OVS) [12] software switch and floodlight controller [13] which support OpenFlow version 1.3 [14]. We modify the source code of the software switches used in Mininet, i.e. OpenVSwitch (OVS), to generate a large volume control traffic required for our experiments. Mininet runs on a Xubuntu VM allocated with four CPU cores and four GB RAM. Each controller instance runs on a Lubuntu VM allocated with one CPU core and one GB RAM. Necessary modifications are made in the OVS source code in order to generate a large volume of OpenFlow packetIn message traffic. With the changes in the OVS code, a tree topology with 131 switches can generate up to 30 Mbps of OpenFlow traffic towards the controller. While generating traffic, each controller uses the default forwarding module, topology manager and link discovery module. Major portion of the controller traffic is generated by the sequential ICMP messages communicated by all the 128 hosts in the topology.

In FOIL, inter-controller messaging is done by ICMP messages shared among the controllers. These inter-controller ICMP messages are detected by TCPDUMP [15] installed in the controller system which notifies the controller about incoming ICMP messages. CPU and memory usage data of the controller devices is collected by the controller by executing shell commands at regular intervals of one second. Whenever the CPU or memory usage of any controller goes beyond the set threshold values, FOIL load balancing procedure is invoked by ICMP message exchange.

All the experiments are performed 20 times and results are averaged to obtain final figures for the experiments where average values are taken.

The experiments are performed for two configurations:

1. In configuration 1, the whole network is controlled by a single controller. The FOIL mechanism is not applicable here.
2. In configuration 2, a FOIL cluster is used to control the whole network. FOIL failure recovery as well as load balancing is applicable in this configuration.

Experiments are conducted to assess FOIL in terms of the parameters such as: memory and CPU usage of controllers, OpenFlow message rate, average migration time taken by the controllers for successful role switching in FOIL during failure recovery or load balancing, response time of the controllers against rate of OpenFlow packetIn messages for different number of controllers operating the network, effect of switch migration on response time and FOIL fault-tolerance in terms of controller response time and TCP packet loss in case of controller failure.

To test bottleneck scenarios in the experiments, for load balancing, we set CPU and memory usage thresholds as 20 and 60%, respectively. These usage values are the overall usage percentages of the system including controller as well as background system processes. It is to note that control traffic volume in the OpenFlow controller does not only include OpenFlow messages but also other control packets such as LLDP, BDDP, etc. These control packets (LLDP, BDDP) are generated for network topology discovery and are also broadcasted periodically for failure detection. Traffic volume due to these control packets stays almost constant for a particular size of the topology. On the other hand, OpenFlow message traffic volume varies largely due to the unpredictable nature of the data traffic that may pass through the network. Therefore, we evaluate all the results for only OpenFlow message traffic that affects the system resources greatly as compared to other less affecting control packets which are excluded in the results.

A. **Memory and CPU usage of controllers**: Fig. 2a shows the average memory and CPU usage of the controllers. This usage percentage is defined as the total usage by the controller process only. As expected, configuration 2 shows much less average CPU and memory usage as compared to configuration 1. It is observed that FOIL reduces CPU utilization significantly as compared to configuration 1 with a single controller.

Fig. 2 **a** Average memory and CPU usage percentage per controller in each configuration, **b** average resource utilization in the controllers in configuration 2 where C1, C2 and C3 denote the three controllers in the FOIL cluster

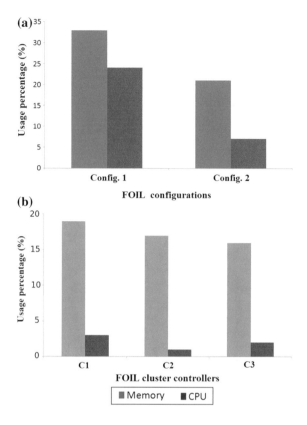

In Fig. 2b, we show the average resource utilization of all the three controllers used in configuration 2. Due to the imposed low thresholds on CPU and memory usage to simulate overload situation and load balancing, the overall memory and CPU usage in all the controllers are low. Results are observed for the first 60 s of the network start-up when OpenFlow traffic volume stays at the peak. FOIL distributes the traffic load evenly across all the controllers in the cluster and thereby avoids over utilization of any controller.

B. **OpenFlow message rate in FOIL**: In order to determine the traffic load on a controller, we observe the controller's OpenFlow message traffic rate which excludes other broadcast control traffic such as LLDP, BDDP, etc. Further, the inter-controller ICMP message traffic is very low as compared to the OpenFlow message traffic and hence we may safely exclude it from this evaluation. In our experiment, most of the OpenFlow message traffic consists of the packetIn messages generated by the OVS. Hence, we plot only the packetIn traffic as in Fig. 3a. The controllers in MASTER and EQUAL role process larger OpenFlow traffic than a controller in SLAVE role. Figure 3a shows the average OpenFlow message traffic rate of the MASTER controller in the two configurations. FOIL in configuration 2 shows lower OpenFlow message traffic load for a MASTER

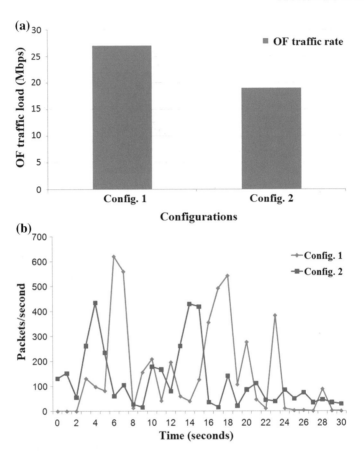

Fig. 3 **a** OpenFlow packetIn message traffic rate for FOIL configurations, **b** variations in OpenFlow message packet rate at the MASTER controller for the two configurations

controller as compared to configuration 1. In configuration 1, the single controller has the highest load of OpenFlow message traffic.

Figure 3b shows the OpenFlow packet rates for 30 s at the MASTER controller in all the three configurations. FOIL configuration 2 shows lower OpenFlow message packet rate due to the sharing of the controller traffic through load balancing. Lower peak value for packets per second makes configuration 2 more stable than others since it means avoidance of sudden bottlenecks at the controller.

C. **Fault-tolerance in FOIL**: This experiment is performed to test the fault-tolerance of FOIL. Here, the test is carried out on configuration 2 with three controllers. Out of the three controllers, two controllers initially with MASTER and EQUAL roles are terminated one after the other at 72nd and 92nd seconds, respectively. The controller initially set in the SLAVE role is observed for the whole period of the experiment. Figure 4 shows the OpenFlow message traffic

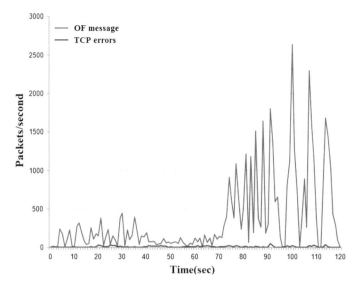

Fig. 4 OpenFlow traffic and TCP error of the remaining controller due to the failure of others in the cluster

rate for 120 s during which two failures are introduced. After the 72nd second, the third controller (initial SLAVE) becomes an EQUAL controller and after the 92nd second it becomes the only controller as well as the MASTER controller. It is to note that till the 71st second, FOIL load balancing takes place and after that failure recovery follows.

Figure 4 also plots the TCP errors in packets per second during the experiment. It is observed that very low and negligible TCP packet loss occurs during switch migration indicating smooth switch and controller connectivity. Due to failures, traffic load increases on the remaining controller but able to handle the traffic with negligible TCP packet losses. Traffic rate for the controller goes high after the second failure when it takes up the MASTER role.

We also observed the time required for switching to the MASTER role during failures in this experiment. Role changing time in this case is found to be between 100 and 150 ms for all the runs of the experiment. The role switching to EQUAL from SLAVE takes less than 50 ms on an average. Thus, FOIL also shows a very fast failure recovery and the ability to cope with multiple controller failures.

D. **Time required for migration**: For measuring the migration time for a switch to a new MASTER, we take the maximum values from all the rounds of an experiment. While the migration time durations were found to vary from 20 to 30 ms and the average was found to be around 27 ms, this shows that the migration time for FOIL is low. It was further observed that in none of the experiments, there was any loss of packets despite migrations which indicates good controller response towards incoming packets. Thus, FOIL provides an efficient switch migration mechanism with very small latency and fast response from the controller.

5 Conclusion

In this paper, we have presented FOIL, a Fault-tolerant OpenFlow multi-controller for the reliable SDN control plane. It provides efficient mechanisms for load balancing among controllers and for failure recovery in the controller cluster. The inter-controller communication is made lightweight with use of ICMP. Performance study of FOIL is carried out in a Mininet-based set-up. From the experimental results, it is seen that FOIL is efficient in terms of failure recovery and load balancing. It achieves a balanced utilization of controller CPU and memory and minimizes per controller traffic load. Further, it introduces a very small delay in switch migration and the controllers display fast response towards the queued OpenFlow packets. The TCP packet loss experienced for switch migration during load balancing is nil and that during controller failure recovery is negligible. Thus, FOIL provides fault-tolerance and thereby reliable control plane for OpenFlow using a distributed approach.

References

1. N. McKeown, T. Anderson, H. Balakrishnan, G. Parulkar, L. Peterson, J. Rexford, S. Shenker, J. Turner, Openflow: enabling innovation in campus networks. ACM SIGCOMM Comput. Commun. Rev. **38**(2), 69–74 (2008)
2. Open Networking Foundation (ONF), OpenFlow Switch Specification, Version 1.2, 2011. https://www.opennetworking.org/images/stories/downloads/sdn-resources/onf-specifications/openflow/openflow-spec-v1.2.pdf. Accessed 30 Jan 2018
3. B. Lantz, B. Heller, N. McKeown, A network in a laptop: rapid prototyping for software-defined networks. Proc. Hotnets. **19**(1–19), 6 (2010)
4. B. Alpern, F. Schneider, Recognizing safety and liveness. Distributed Computing **2**, 117–126 (1987)
5. O. Blial, M.B. Mamoun, R. Benaini, *An overview on SDN architectures with multiple controllers*, vol. 2016 (J. Comput. Netw., Commun, 2016)
6. T. Koponen, M. Casado, N. Gude, J. Stribling, L. Poutievski, M. Zhu, R. Ramanathan, Y. Iwata, H. Inoue, T. Hama, S. Shenker, Onix: a distributed control platform for large-scale production networks, in *Proceedings of the 9th USENIX Conference on Operating Systems Design and Implementation (OSDI'10)* (2010), pp. 1–6
7. A. Dixit, F. Hao, S. Mukherjee, T. Lakshman, R. Kompella, Elasticon: an elastic distributed sdn controller, in *Proceedings of the Tenth ACM/IEEE Symposium on Architectures for Networking and Communications Systems* (2014), pp. 17–28
8. S.H. Yeganeh, Y. Ganjali, Kandoo: a framework for efficient and scalable offloading of control applications, in *Proceedings of the First Workshop on Hot Topic in Software Defined Networks (HotSDN'12)* (2012), pp. 19–24
9. D. Gupta, R. Jahan, Inter-SDN controller communication: using border gateway protocol. http://docplayer.net/5817317-Telecom-white-paper-inter-sdn-controller-communication-using-border-gateway-protocol.html. Accessed 30 Jan 2018

10. D. Li, L. Ruan, L. Xiao, M. Zhu, W. Duan, Y. Zhou, M. Chen, Y. Xia, M. Zhu, High availability for non-stop network controller, in Proceedings of IEEE WoWMoM (2014), pp. 1–5
11. Apache Cassandra. http://cassandra.apache.org/. Accessed 30 Jan 2018
12. Open vSwitch. http://openvswitch.org/. Accessed 30 Jan 2018
13. Project Floodlight. http://www.projectfloodlight.org/floodlight/. Accessed 30 Jan 2018
14. Open Networking Foundation (ONF), OpenFlow Switch Specification, Version 1.3 (2012). http://www.cs.yale.edu/homes/yu-minlan/teach/csci599-fall12/papers/openflow-spec-v1.3.0.pdf. Accessed 30 Jan 2018
15. TCPDUMP. http://www.tcpdump.org/. Accessed 30 Jan 2018

An Efficient Clustering and Routing Algorithm for Wireless Sensor Networks Using GSO and KGMO Techniques

G. R. Asha and Gowrishankar

Abstract Wireless sensor networks (WSNs) play a key role in data transmission based on the locations of sensor nodes (SNs). WSN contains base station (BS) with several SNs, and these SNs are randomly spread across the entire region of monitoring. The BS aggregates the data received from several SNs for meaningful analysis-deployed environment. Energy conservation is the major challenge in the WSN. Since SNs are battery-operated over a period of time, SNs drain their energy in sensing the region of interest and passing on the same to the BS. The consumption of energy depends on the distance between SNs and BS. The SNs are clustered with a certain criteria, and by choosing the cluster head (CH) to aggregate the gathered information by SNs along with determining the optimized path from CH to BS by efficient routing protocol are the innovative techniques in enhancing the lifetime of WSN by optimizing the energy consumption. In this work, an efficient clustering and routing algorithm for WSN using glowworm swarm optimization (GSO) and kinetic gas molecule optimization (KGMO) techniques are proposed. The GSO-KGMO-WSN technique is applied to enhance the lifetime of WSN by effectively reducing the unnecessary data transmission between SNs and BS. This in turn reduces the dead SNs over the period of time which results in enhancing the lifetime of WSN.

Keywords Wireless sensor networks · *K*-means with glowworm swarm optimization clustering algorithm · Kinetic theory of gas molecules routing algorithm

G. R. Asha (✉)
B.M.S. College of Engineering, Research Scholar, Jain University, Bengaluru, India
e-mail: asha.cse@bmsce.ac.in

Gowrishankar
B.M.S. College of Engineering, Bengaluru, India
e-mail: gowrishankar.cse@bmsce.ac.in

© Springer Nature Singapore Pte Ltd. 2020
A. Elçi et al. (eds.), *Smart Computing Paradigms: New Progresses and Challenges*,
Advances in Intelligent Systems and Computing 767,
https://doi.org/10.1007/978-981-13-9680-9_5

1   Introduction

WSNs are one of the popular areas in the networks due to the constant innovations in the field of wireless technology and embedded system. Their application includes monitoring, tracking, event detection, surveillance, disaster management, and preventive maintenance. WSN comprises spatially dispersed autonomous devices called SNs having sensor to monitor environmental or physical conditions with the capability of transmitting the same through wireless technology. A WSN system includes a BS that delivers gathered data to the outside world either by wireless or by wired technology. The work of SNs to send the sensed data to gateway is called as BS. In this process, SNs' battery will be drained due to the transmission of data. Researchers proposed many techniques to conserve energy in WSN. One such technique is clustering; here all SNs are divided into many groups based on certain criteria and elect the CH among these grouped SNs. The SNs in that group send data to CH, and the CH will consolidate these data and transmit to the BS. Since the CHs transmit the data to the BS, significant energy is conserved in routing data to BS. This reduces a lot of energy consumption of individual SN. Traditional schemes for energy-efficient clustering and routing are concerned with only improving the network lifetime. But lots of problems were aroused due to the premature death of CHs and holes. The proposed method uses a class of meta-heuristic algorithms called bio-inspired algorithm which is more suitable to the large-scale application like energy conservation in WSN, where traditional method predominantly fails due to the several conflicting criteria in optimization. The KGMO converges faster and especially suitable for such complex optimization problems [1]. The swarm-based algorithms play a vital role in solving complex real-world problems, so GSO and KGMO are used for optimized cluster formation and efficient data routing.

2   Related Work

In WSN, the SNs' limited battery power can be managed by the clustering technique Heuristic Algorithm for Clustering Hierarchy (HACH) was proposed. The HACH algorithm has two main phases such as sleep scheduling and CH selection. The HACH algorithm balances and minimizes energy utilization by choosing distributed nodes with high energy as CH to increase network lifetime, but fails to explain the concept of heterogeneity in terms of energy levels at WSN [2].

The key parameters that play a significant role in lifetime of WSN are energy and computational feasibility where the SNs are resource constrained. Energy consumption primarily depends on age of WSN and the distance from BS and SNs. The distance results in unbalanced usage of energy. The formation of clustering and intelligent routing concept was introduced through genetic algorithm (GA). The low-energy adaptive clustering hierarchy (LEACH) is one of the early clustering algorithms for SNs in WSN. The LEACH does not support the random selection of the CHs. The several innovative techniques such as updating of CH at the end of round of data transmission were introduced to improve the lifetime of WSN, but fail to address the issue of end-to-end packet delay [3].

Since WSN functions in the energy-constrained environment, optimized data routing will play a major role. In improved routing algorithm for WSNs using ant colony optimization (ACO), the issue of optimized routing is addressed. The ACO algorithm was compared with the energy-efficient ant-based routing (EEABR) and leach-ant algorithm to prove the minimum energy consumption, but the major issue of packet delivery ratio was not addressed [4].

The ring communication topology was the early and efficient routing technique in wired network, and the same concept was extended for a heterogeneous WSN to conserve energy in WSN. The energy-efficient heterogeneous ring clustering (E2HRC) algorithm guarantees better average energy consumption. The algorithm is compared with the original IPv6 routing protocol, Low-Power and Lossy networks (RPL). The issue related to the lifetime of the network was not discussed [5].

The energy consumption and load balancing problems were addressed by introducing the improved routing protocol (IRPL). Here the routing topology control model has been segmented into rings of equal area for better load balancing and balanced energy conservation. The issue pertaining to the packet delay was not addressed [6].

The concept of clustering for data aggregation among SNs and efficient routing data between CHs for energy conservation has not been addressed comprehensively by any of these techniques. The GSO-KGMO-WSN methodology addressed the issue of clustering and routing issue as a cohesive problem through bio-inspired GSO and KGMO concepts.

3   GSO-KGMO-WSN Methodology

The aim of the GSO-KGMO-WSN method is to provide the energy-efficient clustering and optimized path to transmit the data packet from SNs to the BS over the network. In this method, after deploying SNs, initial clusters are formed using *K*-means algorithm, clusters are optimized, and CHs are elected through GSO algorithm. The data aggregated by the clustering is transmitted to the BS trough the CH by multi-hop routing involving various CHs. The routing path is optimized through KGMO algorithm. The energy-efficient process of clustering and routing is illustrated schematically in Fig. 1.

3.1   Creation of WSN

Sensor network is a collection of large number of heterogeneous SNs which are having different processing capabilities, energy level, sensors, and transmission

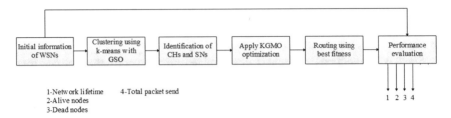

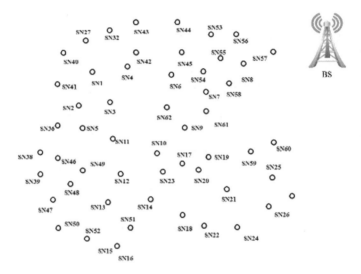

Fig. 1 Block diagram of GSO-KGMO-WSN method

Fig. 2 Random sensor deployment

range. These SNs are randomly deployed in a network based on the requirement of the application. The inter-SN distance plays a vital role in determining the lifetime of the network. The position, energy level, type, and identifications (IDs) are communicated to the BS to provide the deployment information of the SNs. The creation of WSN is shown in Fig. 2.

3.2 Identification of Cluster Head

In the process of creation of WSN, the SNs send the complete information to the BS. The BS initially forms the K-clusters of SNs by K-means clustering technique. The clusters are formed based on the energy level of SN and the distance between SNs. The K-clusters are formed, and a CH is elected in each cluster. The clusters formed are further subjected to GSO algorithm for optimized clustering. Here, SNs are considered as glowworm (GW), and each GW contains luminance quantity called

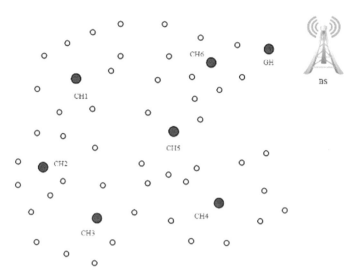

Fig. 3 Cluster formation

as luciferin intensity. The intensity of luciferin is determined by residual energy of SN, its distance with CH, and total number of SNs associated with the CH. In the process, each GW is attracted towards neighbour GW based on the luciferin intensity. The SN with highest luciferin intensity will be elected as a CH. The cluster formation is illustrated schematically in Fig. 3.

3.3 Routing Process

After identifying the CHs, the KGMO is used to optimize the routing policy between the CHs and BS. The fitness function of SN determines the intermediate SN to be preferred during the routing process. The fitness function is validated at each iteration process to choose the candidate SN for routing. In this process, the SN with less residual energy will be automatically eliminated in the optimized route. The fitness function of KGMO routing process is determined by the residual energy of each SN, the inter-cluster distance, and the density of route.

$$f = \text{Fitness function} = e + d + n \tag{1}$$

where e is the residual energy of each SN, d is the inter-cluster distance, and n is the number of hops. The fitness function determines the optimized route.

In this process, the CH with the less residual energy will be automatically eliminated from the optimized path at each round. When the SN reaches a threshold value

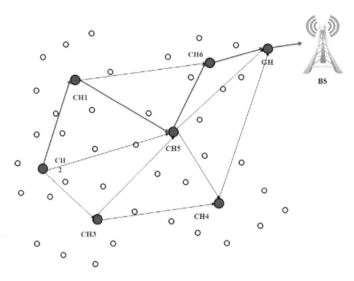

Fig. 4 Data transition in a network

of energy, the SN will be declared as dead SN. The optimized route is illustrated schematically in Fig. 4.

4 Performance Evaluation

In order to determine the optimized clustering and efficient routing in WSN, the parameters are residual energy, Received Signal Strength (RSS), distance and type of sensors. The residual energy and distance are the factors that affect the lifetime of the WSN. Lifetime/rounds, total no. of active nodes, the dormant nodes, and packet transfer ratio are the metrics for the performance evaluation. The cluster quality determines the energy consumption of the WSN. The efficient routing enhance the lifetime of the network. In general, better the cluster quality and the shortest path determines the efficacy of algorithm.

The algorithm GSO-KGMO-WSN governed by the fitness function and the parameters of fitness function comprises the residual energy, the intra-cluster distance, and the number of hops. The major parameters of the fitness function are also the factors of the performance evaluation. The proposed algorithm is compared with popular energy-efficient algorithm PSO-PSO-WSN and previously reported work PSO-GSO-WSN [7].

The outcome of the performance evaluation depicts that having a better control on the fitness function will lead to efficient load balancing among the CHs in optimal routing and energy conservation in forming the compact clusters which in turn lead to the improvement of network lifetime.

The packet delivery ratio is one of the important performance measures of routing algorithms in WSN. In the course of operation of WSN, the CH may drop the packet due to the fall of energy below the threshold value. The routing algorithm should dynamically adapt to such situation by eliminating those CHs from the critical path of routing. In GSO-KGMO-WSN methodology, one of the key parameters of fitness function is residual energy of SN. At each round, residual energy of SN is considered in deciding the optimized route from CH to BS. Those CHs having the energy level lesser than the threshold value will be automatically avoided in computing the route. Hence, the packet delivery ratio of GSO-KGMO-WSN will be much better than PSO-PSO-WSN and PSO-GSO-WSN algorithms.

5 Experimental Set-up

To demonstrate the better performance of clustering and routing in WSNs, the GSO-KGMO-WSN method was implemented in MATLAB version R2015b. The completed work was done using the i7 system with 8 GB RAM. The two algorithms are used such as K-means with GSO and KGMO to achieve the best clustering and routing for WSNs. Table 1 contains simulation parameters that are used in common for both algorithms, and Tables 2 and 3 show the parameters used in the GSO and KGMO algorithms. These three tables are used as the input for this GSO-KGMO-WSN methodology.

Table 1 Simulation parameters

Parameter	Value
Area	$250 * 250 \text{ m}^2$
Sensor nodes	300
Gateways	30
Initial energy of sensor nodes	0.5 J
Number of simulation iterations	350
Communication range	150 nm
Eelec	50 PJ/bit
εfs	10 PJ/bit/m^2
εmp	0.0013 PJ/bit/m^4
$d0$	87.0 m
EDA	5 nJ/bit
Packet size	4000 bits
Message size	200 bits

Table 2 GSO parameters

Parameter	Value
ρ	0.4
γ	0.6
β	0.08
nt	5
s	0.03
Initial luciferin level	5

Table 3 KGMO parameters

Parameter	Value
Del value	0.1
$C1$	1
$C2$	3
T	1

6   Results and Discussion

The GSO-KGMO-WSN was experimented with 300 sensor nodes and 30 gateways for generating the effective clustering and routing process in WSNs. Assume each sensor node has the initial energy up to 0.5 J. GSO-KGMO-WSN algorithm is verified and illustrates the experimental results for both routing and clustering. The coverage area of the entire network is 250 * 250 m^2, and the position of the base station is 125,325. For knowing the effective results of the GSO-KGMO-WSN method, it is compared with two existing methodologies (i) PSO-PSO-WSN and (ii) PSO-GSO-WSN. To execute this methodology, the values of the gamma, beta, step size, luciferin decay constant, luciferin enhancement constant, and del value are initially taken as per Tables 2 and 3.

Figures 5, 6, and 7 present the comparison of GSO-KGMO-WSN with two existing methodologies: (i) PSO-PSO-WSN and (ii) PSO-GSO-WSN. Figure 2 displays the alive nodes which are greatly increased compared to two methodologies such as (i) PSO-PSO-WSN and (ii) PSO-GSO-WSN. The network lifetime and the number of transmissions are increased by the maximization of alive nodes.

Figure 3 shows the results of number of dead nodes, compared to two other methodologies such as (i) PSO-PSO-WSN and (ii) PSO-GSO-WSN; the dead nodes are highly decreased, and it also increases the lifetime of the network. To avoid the node to be dead, the residual energy of each SN is to be maintained for each iteration. It is carried out for eliminating the failure of nodes inside the transmission. In PSO-PSO-WSN and PSO-GSO-WSN methods, the SNs drain their energy after some transmissions; it causes the dead nodes to be more.

Figure 4 displays the results of the total packet sent, compared to an existing method the GSO-KGMO-WSN method transmits the high amount of information

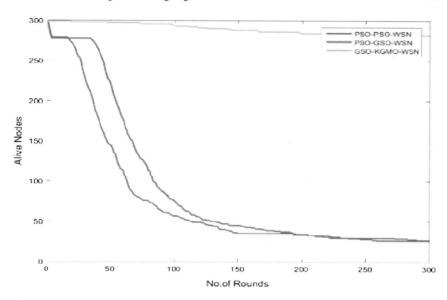

Fig. 5 Alive nodes versus no. of rounds

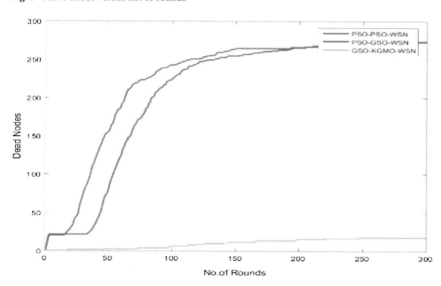

Fig. 6 Dead nodes versus no. of rounds

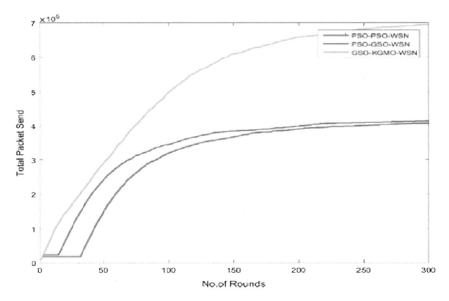

Fig. 7 Total packet sent versus no. of rounds

at the same time. The PSO-PSO-WSN and PSO-GSO-WSN algorithms have high end-to-end delay so that the amount of transmission becomes small.

By increasing the no. of alive nodes and decreasing the number of dead nodes, we directly see that the residual energy is maintained for longer duration compared to other methods. This results in conserving more energy and thus maximizing the network life.

7 Conclusion

In this paper, the GSO and KGMO algorithms are introduced; this has improved the performance of clustering and routing in WSNs. Using K-means clustering, centroids are identified which are further optimized by GSO algorithm. The elected CHs are subjected to the KGMO methodology, whose fitness function is based on the residual energy, distance, and the number of hops in a network. The optimal route is established only based on this fitness function. By assigning a minimum threshold for each sensor node, the failure of nodes is eliminated in routing at each iteration. The failure of nodes creates the packet drop among WSN, and it degrades the performance of desired the network. Network lifetime is increased by minimizing the energy consumption among the network and the alive nodes, which is present in the network is increased as well as the amount of packets received by the BS also increased. From obtained results, we conclude that the GSO-KGMO-WSN methodology guarantees better efficiency compared to the other two algorithms: PSO-PSO-WSN and PSO-GSO-WSN.

Acknowledgements The work reported in this paper is supported by the college through the TECHNICAL EDUCATION QUALITY IMPROVEMENT PROGRAMME [TEQIP-III] of the MHRD, Government of India.

References

1. S. Moein, R. Logeswaran, KGMO: a swarm optimization algorithm based on the kinetic energy of gas molecules. Inf. Sci., 127–144 (2014)
2. M.O. Oladimeji, M. Turkey, S. Dudley, HACH: heuristic algorithm for clustering hierarchy protocol in wireless sensor networks. Appl. Soft Comput. **55**, 452–461 (2017)
3. P. Nayak, B. Vathasavai, Genetic algorithm based clustering approach for wireless sensor network to optimize routing techniques, in *7th International Conference on Cloud Computing, Data Science & Engineering-Confluence*. (2017), pp. 373–380
4. Y. Sun, W. Dong, Y. Chen, An improved routing algorithm based on ant colony optimization in wireless sensor networks. IEEE Commun. Lett. **21**, 1317–1320 (2017)
5. W. Zhang, L. Li, G. Han, L. Zhang, E2HRC: an energy-efficient heterogeneous ring clustering routing protocol for wireless sensor networks. IEEE Access **5**, 1702–1713 (2017)
6. W. Zhang, G. Han, Y. Feng, J. Lloret, IRPL: an energy efficient routing protocol for wireless sensor networks. J. Syst. Architect. **75**, 35–49 (2017)
7. G.R. Asha, Gowrishankar, An energy aware routing mechanism in WSNs using PSO and GSO algorithms, in *5th International Conference on Signal Processing and Integrated Networks* (2018)

Cloud Computing, Distributed Systems, Social Networks, and Applications

Survey on Security Issues in Mobile Cloud Computing and Preventive Measures

Rahul Neware, Kalyani Ulabhaje, Gaurav Karemore, Hiteshree Lokhande and Vijay Dandige

Abstract Mobile Cloud Computing (MCC) is a recent technology used by various people worldwide. In 2015, more than 240 million users used mobile cloud computing which earns a profit of \$5.2 billion for service providers. MCC is a combination of mobile computing and cloud computing that presents various challenges like network access, elasticity, management, availability, security, privacy etc. Here, the security issues involved in both mobile computing and cloud computing, such as data security, virtualisation security, partitioning security, mobile cloud application security and mobile device security are considered extremely important. This paper presents a detailed study of security issues in mobile cloud computing and enumerates their preventive measures.

Keywords Mobile computing · Cloud computing · Security · Virtualisation · Privacy · Authentication · Storage

1 Introduction

Mobile cloud computing (MCC) is a recent technology used by various people in their daily lives. It is estimated that roughly 1.5 billion smartphone users and 640 million

R. Neware (✉) · K. Ulabhaje · G. Karemore · H. Lokhande
Computer Science and Engineering Department, G H Raisoni College of Engineering, Nagpur, Maharashtra, India
e-mail: neware_rahul.ghrcemtechcse@raisoni.net

K. Ulabhaje
e-mail: ulabhaje_kalyani.ghrcemtechcse@raisoni.net

G. Karemore
e-mail: Karemore_gaurav.ghrcemtechcse@raisoni.net

H. Lokhande
e-mail: lokhande_hiteshree.ghrcemtechcse@raisoni.net

V. Dandige
Department of English, MNLU, Nagpur, Maharashtra, India
e-mail: vijaydandige@gmail.com

© Springer Nature Singapore Pte Ltd. 2020
A. Elçi et al. (eds.), *Smart Computing Paradigms: New Progresses and Challenges*,
Advances in Intelligent Systems and Computing 767,
https://doi.org/10.1007/978-981-13-9680-9_6

tablet users in the world use mobile cloud computing. Mobile cloud computing (MCC) is the blend of cloud computing, mobile computing and wireless networks that brings rich computational resources to mobile users, network operators, as well as cloud computing providers [1]. The simplified definition of mobile cloud computing is: distributed computing is characterised as the pattern in which resources are given to a customer on an on-demand premise, for the most part by methods through the web [1]. Mobile cloud computing uses infrastructure as a service platform of the cloud for storage and processing, and cloud-based applications move the computational power and information storage into the cloud [2]. MCC is a rich mobile computing technology that leverages unified elastic resources of varied clouds and network technologies towards unrestricted functionality, storage and mobility, and serves a multitude of mobile devices anywhere anytime through the Internet regardless of heterogeneous environments and platforms based on the pay-as-you-use principle [3]. Mobile cloud computing is used by the user using various browsers available like Chrome, Firefox. UC Browser, etc. [4].

Mobile cloud computing is one of the quickest developing segments of the cloud computing worldview. Apple and Google are the two playing the main role in the development of mobile cloud computing. By 2016, 60% of mobile development industries have used cloud services as pay-as-use that reduces resource deficiency and makes devices compatible to use cloud services in mobile devices.

2 Mobile Cloud Service Models

The concept of mobile cloud computing is categorised in different service models. Some of the prominent models of mobile cloud services are as follows.

2.1 Mobile Cloud Infrastructure as a Service (MIaaS)

This service model provides the cloud environment and storage facility for the mobile user. It is like the infrastructure as a service model of cloud which provides all the infrastructure for cloud. The example of MIaaS is Apple iCloud: it is Apple's own cloud-based storage system and initially, it gives 5 GB free storage. The other examples are Amazon Cloud, Dropbox, Google Drive, Microsoft OneDrive, etc.

2.2 Mobile Network as a Service (MNaaS)

This service model offers network infrastructure to users for creating a network. In other words, we can say that mobile network as a series is used to create a virtual network and for connecting mobiles with servers. The example of MNaaS is OpenStsck:

it is used to create virtual networks. The other examples are CoreCluster, OpenVZ, SmartOS, etc.

2.3 Mobile Data as a Service (MDaaS)

This service model provides database service so that mobile cloud users can perform data management and other operations to their data. Example: CloudDB-Cloud-based database made for mobile cloud computing. Oracle's mobile cloud data as a service [5].

2.4 Mobile Multimedia as a Service (MMaaS)

This service model offers a platform to access or run the multimedia in the cloud environment, like playing high-memory capacity required games, playing high-definition videos, etc. [6].

2.5 Mobile App as a Service (MAppaaS)

This service model provides a platform to users for executing the app, and the using app also manages the apps using the wireless network. Examples: Apple App Store, Google Play Store, etc.

2.6 Mobile Community as a Service (MCaaS)

This service model offers the facility to mobile users to create a community network or social network and to manage all such networks and get the services needed for them [7].

3 Generalised Security Requirements

International Telecommunication Union (ITU) and US National Security Agency [8, 9] have defined and laid down certain generalised security requirements of mobile cloud computing. They are as follows.

3.1   Confidentiality

Confidentiality is a fundamental requirement because mobile users' data are processed through the public network and is also stored in public servers. So there is a
high chance of unauthorised access to mobile users' data, owing to which the issue
of confidentiality is a big challenge to mobile cloud service providers.

3.2   Availability

Availability means cloud service is always available for users 24/7 when they need the
service. There are various attacks that affect availability, but mobile cloud computing
service providers need to prevent them and always ensure the service is available for
mobile users.

3.3   Authentication and Access Control

This means identifying the valid user of the system by some login patterns or any other
mechanism called authentication. Giving access to limited resources to authenticate
users of the system as they want to do some task is called access control. Actions
performed by users like reading, writing, updating, erasing data, etc. are all controlled
in access control.

3.4   Integrity

Integrity means prevention of data loss or data modification while transmitting it
through the public network. Integrity deals with consistency and accuracy of user
data.

3.5   Privacy

Privacy is the security of mobile user's personal data while communicating in the
cloud, achieved through confidentiality, integrity and authentication.

4 Challenges in Mobile Cloud Computing

Mobile cloud computing is a service of cloud computing used in smartphones or in tablets. Mobile computing and cloud computing combine together to form mobile cloud computing and give services of cloud to mobile computing users like on-demand self-service, resource pooling measured services, elasticity, broad network access [10, 11]. Mobile cloud computing uses wireless communication technology to communicate between mobile and cloud [12]. Owing to the combination of mobile computing and cloud computing and use of wireless communication, we face many challenges in mobile cloud computing, such as limited resources for mobile devices, stability challenge occurring due to limitation of wireless network, cost of network access going high various times in mobile cloud computing, elasticity challenge, security and privacy challenge, bandwidth of channel, energy efficiency, quality of service, etc. [13, 41].

This paper is divided into four sections. The introduction section throws light on the journey of mobile cloud computing, its various definitions, statistics, service models and security requirements of mobile cloud computing and challenges it faces. The second section discusses the security issues of mobile cloud computing. The third section presents preventive measures relating to security issues, and the fourth section gives the conclusion (Fig. 1).

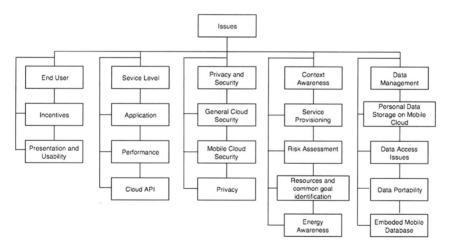

Fig. 1 Issues in mobile cloud computing

5   Security Issues in Mobile Cloud Computing

5.1   Data Security Issues

In mobile cloud computing mobile user's data are available and stored in the cloud and the processing of that data is also done in IaaS of the cloud. Many attacks are executed on data of mobile cloud computing like data loss, data breach, data recovery from damage, data locality, data correctness, etc. In data loss, user's data are mossed while performing any computational task; for example, while transmitting data through public network. In data breaches, an authorised user's data are accessed by an unauthorised person by injecting into the cloud or by getting it using any unwanted activity. In data recovery from damage issues, a user should get valid data of his own while recovering due to damage of system or mobile device. Cloud stores the data in any data centre; so the location of that data not known to anyone. So the challenge is that the user should know where his important data are stored. Data management is done in the service providers' premises and they need to maintain confidentiality and integrity [14].

5.2   Virtualisation Security Issues

Cloud services are provided to mobile users using virtualisation. A virtual machine of mobile is re-installed in the cloud, which is called as mobile clone, and this cloud-based virtual machine does all the processing. The main advantage of using a virtual machine is that it creates instances of various machines and this is achieved through the hypervisor. But the challenges to the virtual machine used in cloud computing are unauthorised access to the main machine through virtual machine, root attack, VM to VM attack, communication in virtualisation and confidentiality of data while being processed through hypervisor [15, 16].

5.3   Offloading Security Issues

Offloading means transformation of the task to an external platform. Mobile cloud computing requires wireless network for offloading in the cloud, but precisely because of this, unauthorised access of data is possible during offloading. The main issue in offloading is availability which happens because of the jamming of the mobile device while the offloading is taking place. Also, while offloading of data if it contains any malicious content, then it affects the confidentiality and privacy of the mobile user.

5.4 Mobile Cloud Applications Security Challenges

Various mobile cloud applications are affected by various malware, worms, Trojan horse, botnet, etc., which in turn affect the confidentiality and integrity aspects. These malware are run in mobile devices and bind themselves in the application and mutate, which cause very serious issues to mobile cloud computing [17–19].

5.5 Mobile Device Security Issues

This is the most ubiquitous issue in mobile cloud computing occurring due to theft or loss of the mobile device. Here, the main loss is of the user's data. If the attacker gets access into any mobile device then unauthorised access to data and application occurs. The device can also be used to do some unwanted tasks like botnet: to carry out DoS or DDoS attack through mobile device [20]. A new attack related to power consumption is carried out on mobile devices when the device is connected to the wireless network; then its power consumption increases to discharge device battery fast [21]. Mostly, a mobile device stores user's personal information into internal storage; but when the user uses mobile cloud computing then all data are synchronised to cloud and there the security of user's personal data becomes insecure. In mobile computing, malware and viruses constitute the very old methods of attacks but they are effective and work in mobile devices because the mobile operating system is neither secure nor strong (Fig. 2).

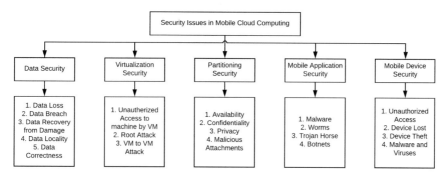

Fig. 2 Security issues in mobile cloud computing

6 Preventive Measures to Current Security Issues of MCC

6.1 Preventive Methods for Data Security Issues

Mollah et al. [22] have given data sharing and searching technique for mobile devices for sharing and searching data in the cloud securely through public and private key encryption and digital signature. Sookhak et al. [23] introduced the remote data audition method which verifies and stores data integrity in the cloud. He also developed a divide and conquer table (DCT), which updates at the block level. Odelu et al. [24] proposed a technique to access control and outsourcing computational process from cloud to mobile. Here, two schemes are developed: Secure and lightweight CP-ABE (SL-CP-ABE) and CP-ABE-Constant size type text and secret key (CP-ABE-CSCTSK).

Li et al. [25] developed secure accessing in cloud platform 2SBM (Inter-crossed Secure Big Multimedia Model), which is based on ontology access recognition and matching algorithm. Alqahtani and Kouadri-Mostefaou [26] designed the framework based on distributed multi-cloud storage, encryption and data compression, in which data are divided into segments, then the encryption of segments takes place and then these encrypted segments are compressed.

6.2 Preventive Methods for Virtualisation Security Issues

Liang et al. [27] have given a technique to secure virtual machine deployment, which uses mandatory access control technique to control resources and gives powerful isolation from guest virtual machines. Hao et al. [28] invented a new SMOC technique which gives permission to make a copy of operating system and application of mobile to a virtual machine on the cloud for data security. Paladi et al. [29] developed cloud infrastructure which includes virtual machine launching and data protection protocol. Virtual machine launching is used before a guest virtual machine and data protocol ensure confidentiality using the cryptographic technique.

Jin et al. [30] developed (hardware-assisted secure virtual machine (H-SVM) technique which secures guest virtual machine from infected hypervisor by memory virtualisation. This technique is very useful and it is less vulnerable. Vaezpour et al. [31] developed a SWAP technique for phone clone. This technique is based on two other techniques: first, securing mobile clone to lessen the threats for data leakage from the virtual machine and second, migration of clone when the threatened virtual machine is at high-level risk.

6.3 Preventive Methods for Offloading Security Issues

Duan et al. [32] have given application offloading technique in which users' private information is kept within the mobile while offloading is being performed. Owing to this, unauthorised access and integrity security problems do not occur. This technique preserves privacy and also saves energy. Al-Mutawa and Mishra [33] used data partitioning technique to prevent exposure of user's personal information during offloading. This technique consists of three steps: in the first, data are divided into sensitive and non-sensitive segments, in the second, sensitive data processing takes place on the device and in the third, non-sensitive data processing takes place in cloud.

Saab et al. [34] have given mobile application offloading technique that comprises profiling, decision-making and offloading engine. Profiling and decision-making are used for dynamic partitioning to reduce power consumption and decrease security issues. Offloading engine is used to offload app to cloud for processing. Khan et al. [35] proposed cloud manager-based re-encryption scheme (CMReS) cryptographic method which protects offloading. This method uses encryption, decryption and re-encryption of data for more security while offloading and, additionally, it is under the control of client organisation.

6.4 Preventive Methods for Mobile Cloud Applications Security Issues

Tang et al. [36] proposed application program interface (API) model consisting of three factors: first, authentication with user registration with storing passwords, second, security mechanism like encryption, decryption, digital signature and third, API with backend services. Popa et al. [37] secure mobile cloud (SMC) developed an application to confirm the security of data communication in the cloud as well as in mobile device. It measures the integrity of application when interacting with a mobile device. In application integrity, first verification of application takes place, and then its signature is matched with the original application signature for finding any attachment in it. In this application, six different managers are used: mobile manager, mobile security manager, cloud security manager, optimisation manager, application manager and policy manager.

6.5 Preventive Methods for Mobile Device Security Issues

Sitová et al. [38] have given a technique for authentication of the user by using biometric features. In this technique, biometric includes the hand movement of a user grasping the mobile which generates patterns and those patterns are used for identification of authorised and unauthorised users. In [39] authors discussed Google device policy application. This application is very much useful when the mobile device is stolen or lost user can clean his data online and also enables device token (unique key) which ensures the notification security. Imgraben et al. [40] have given the approach of OpenFlow in which OpenFlow switch is integrated with mobile to do the job of redirection, while communication of mobile and all cloud data is passed through OpenFlow, so that the data are secure while being transmitted.

7 Conclusion

In the first section of this paper, mobile cloud computing is explained in detail, which includes definition, history and security introduction about MMC. Various security issues are available for mobile cloud computing. However, the issues discussed in this paper are basic and very important. The prevention measures enumerated are very recent solutions to security issues of mobile cloud computing.

References

1. http://www.cse.wustl.edu/~jain/cse574–10/ftp/cloud/index.html,   https://en.wikipedia.org/wiki/Mobile_cloud_computing
2. A Report of Worldwide Smartphone Markets: 2011 to 2015, May 2011: http://www.researchandmarkets.com/research/7a1189/worldwide_smartphone
3. Z. Sanaei, S. Abolfazli, A. Gani, M. Shiraz, in *SAMI: Service-Based Arbitrated Multi-tier Infrastructure for Mobile Cloud Computing*. 1st IEEE International Conference on Communications in China Workshops (ICCC) (2012), pp. 14–19
4. R. Neware, Computer forensics for private web browsing of UC browser. IOSR J. Comput. Eng. (IOSR-JCE) **19**(4), 56–60 (2017)
5. L. Lei, S. Sengupta, T. Pattanaik, J. Gao, in *MCloudDB: A Mobile Cloud Database Service Framework*. 3rd IEEE International Conference on Mobile Cloud Computing, Services, and Engineering (MobileCloud) (2015), pp. 6–15
6. W. Zhu, C. Luo, J. Wang, S. Li, Multimedia cloud computing. Sig. Process. Mag. IEEE **28**, 59–69 (2011)
7. D. Kovachev, D. Renzel, R. Klamma, Y. Cao, in *Mobile Community Cloud Computing: Emerges and Evolves*. Eleventh International Conference on Mobile Data Management (MDM) (2010), pp. 393–395
8. Security in Telecommunications and Information Technology: An Overview of Issues and the Deployment of Existing ITU-T Recommendations for Secure Telecommunications (2016), https://www.itu.int/itudoc/itu-t/85097.pdf
9. US National Security Agency: Information Assurance (2016), http://www.nsa.gov/ia/ia_at_nsa/index.shtml.online
10. M.B. Mollah, K.R. Islam, S.S. Islam, in *Next Generation of Computing Through Cloud Computing Technology*. 25th IEEE Canadian Conference on Electrical and Computer Engineering (CCECE) (2012), pp. 1–6
11. R. Buyya, C.S. Yeo, S. Venugopal, J. Broberg, I. Brandic, Cloud computing and emerging IT platforms: vision, hype, and reality for delivering computing as the 5th utility. Future Gener. Comput. Syst. **25**, 599–616 (2009)

12. N. Fernando, S.W. Loke, W. Rahayu, Mobile cloud computing: a survey. Future Gener. Comput. Syst. **29**, 84–106 (2013)

13. R. Neware, A. Khan, in *Cloud Computing Digital Forensic Challenges*. 2nd International conference on Electronics, Communication and Aerospace Technology (ICECA 2018), Coimbatore, India (2018)

14. R. Neware, Recent threats to cloud computing data and its prevention measures. Int. J. Eng. Sci. Res. Technol. **6**(11), 234–238 (n.d.)

15. R. Sharma, S. Kumar, M.C. Trivedi, Mobile, in *Cloud Computing: Bridging the Gap between Cloud and Mobile Devices*. 5th International Conference on Computational Intelligence and Communication Networks (CICN) (2013), pp. 553–555

16. R. Neware, N. Walde, Survey on security issues of fog computing. Int. J. Innovative Res. Comput. Commun. Eng. **5**(10), 15731–15736 (2017). https://doi.org/10.15680/IJIRCCE.2017. 0510009

17. V. Prokhorenko, K.-K.R. Choo, H. Ashman, Web application protection techniques: a taxonomy. J. Netw. Comput. Appl. **60**, 95–112 (2016)

18. J. Peng, K.-K.R. Choo, H. Ashman, User profiling in intrusion detection: a review. J. Netw. Comput. Appl. **72**, 14–27 (2016)

19. D. Quick, K.-K. R. Choo, Pervasive social networking forensics: intelligence and evidence from mobile device extracts. J. Netw. Comput. Appl. (2016)

20. L. Liu, X. Zhang, G. Yan, S. Chen, in *Exploitation and Threat Analysis of Open Mobile Devices*. Proceedings of the 5th ACM/IEEE Symposium on Architectures for Networking and Communications Systems (2009), pp. 20–29

21. R. Racic, D. Ma, H. Chen, Exploiting MMS vulnerabilities to stealthily exhaust mobile phone's battery. Securecomm Workshops **2006**, 1–10 (2006)

22. M.B. Mollah, M.A.K. Azad, A. Vasilakos, Secure data sharing and searching at the edge of cloud-assisted internet of things. IEEE Cloud Comput. **4**, 34–42 (2017)

23. M. Sookhak, A. Gani, M.K. Khan, R. Buyya, Dynamic remote data auditing for securing big data storage in cloud computing. Inf. Sci. **380**, 101–116 (2017)

24. V. Odelu, A.K. Das, Y.S. Rao, S. Kumari, M.K. Khan, K.-K. R. Choo, Pairing-based CP-ABE with constant-size ciphertexts and secret keys for cloud environment. Comput. Stand. Interfaces (2016)

25. Y. Li, K. Gai, Z. Ming, H. Zhao, M. Qiu, Intercrossed access controls for secure financial services on multimedia big data in cloud systems. ACM Trans. Multimedia Comput. Commun. Appl. (TOMM) **12**, 67 (2016)

26. H.S. Alqahtani, G. Kouadri-Mostefaou, in *Multi-clouds Mobile Computing for the Secure Storage of Data*. Proceedings of the 2014 IEEE/ACM 7th International Conference on Utility and Cloud Computing (2014), pp. 495–496

27. H. Liang, C. Han, D. Zhang, D. Wu, A lightweight security isolation approach for virtual machines deployment, in *Information Security and Cryptology* (2014), pp. 516–529

28. Z. Hao, Y. Tang, Y. Zhang, E. Novak, N. Carter, Q. Li, in *SMOC: A Secure Mobile Cloud Computing Platform*. 2015 IEEE Conference on Computer Communications (INFOCOM) (2015), pp. 2668–2676

29. N. Paladi, C. Gehrmann, A. Michalas, Providing user security guarantees in public infrastructure clouds. IEEE Trans. Cloud Comput. (2016)

30. S. Jin, J. Ahn, J. Seol, S. Cha, J. Huh, S. Maeng, H-SVM: hardware-assisted secure virtual machines under a vulnerable hypervisor. IEEE Trans. Comput. **64**, 2833–2846 (2015)

31. S.Y. Vaezpour, R. Zhang, K. Wu, J. Wang, G.C. Shoja, A new approach to mitigating security risks of phone clone co-location over mobile clouds. J. Netw. Comput. Appl. (2016)

32. Y. Duan, M. Zhang, H. Yin, Y. Tang, in *Privacy-Preserving Offloading of Mobile App to the Public Cloud*. 7th USENIX Workshop on Hot Topics in Cloud Computing (HotCloud 15) (2015)

33. M. Al-Mutawa, S. Mishra, in *Data Partitioning: An Approach to Preserving Data Privacy in Computation Offload in Pervasive Computing Systems*. Proceedings of the 10th ACM symposium on QoS and security for wireless and mobile networks (2014), pp. 51–60

34. S.A. Saab, F. Saab, A. Kayssi, A. Chehab, I.H. Elhajj, Partial mobile application offloading to the cloud for energy-efficiency with security measures. Sustain. Comput. Inf. Syst. **8**, 38–46 (2015)
35. A.N. Khan, M.M. Kiah, M. Ali, S. Shamshirband, A cloud-manager-based re-encryption scheme for mobile users in cloud environment: a hybrid approach. J. Grid Comput. **13**, 651–675 (2015)
36. S.L. Tang, L. Ouyang, W.T. Tsai, in *Multi-factor web API Security for Securing Mobile Cloud*. 2015 12th International Conference on Fuzzy Systems and Knowledge Discovery (FSKD) (2015), pp. 2163–2168
37. D. Popa, M. Cremene, M. Borda, K. Boudaoud, in *A Security Framework for Mobile Cloud Applications*. 2013 11thRoedunet International Conference (RoEduNet) (2013), pp. 1–4
38. Z. Sitová, J. Šeděnka, Q. Yang, G. Peng, G. Zhou, P. Gasti et al., HMOG: new behavioral biometric features for continuous authentication of smartphone users. IEEE Trans. Inf. Forensics Secur. **11**, 877–892 (2016)
39. Device Policy for Android: Overview for Users (2016), www.google.com/support/mobile/bin/answer.py?hl=en&answer=190930
40. J. Imgraben, A. Engelbrecht, K.-K.R. Choo, Always connected, but are smart mobile users getting more security savvy? A survey of smart mobile device users. Behav. Inf. Technol. **33**, 1347–1360 (2014)
41. R. Neware, Internal intrusion Detection for Data Theft and Data Modification using Data Mining. Int. J. Sci. Res. (IJSR) **6**(8), 2176–2178 (2017). https://www.ijsr.net/archive/v6i8/v6i8.php

A Robust Lightweight ECC-Based Three-Way Authentication Scheme for IoT in Cloud

Sayantan Chatterjee and Shefalika Ghosh Samaddar

Abstract Internet of things is an evolving technology which connects multiple embedded devices with a remote server over Internet. Due to limited capacity of embedded devices, it is important to delegate resources from third-party platform. Lots of research is going on to connect IoT devices to a wide resource pool such as cloud. This integration of IoT with cloud services has enormous possibility for future as resource-intensive processes can be delegated to cloud platform instead of executing in IoT node. This paper proposes a lightweight three-way authentication scheme for IoT in cloud where mutual authentication between IoT node and user's smart device is performed by remote IoT gateway in cloud. The proposed scheme uses a three-factor user authentication to prevent device theft attack and ECC-based communication protocols to ensure less computation and communication overhead. Moreover, the proposed scheme is analysed to show that it is secure against existing relevant cryptographic attacks.

Keywords Internet of things (IoT) · Cloud · IoT device · User's smart device · IoT gateway · Elliptic curve cryptography (ECC) · Mutual authentication · Session key negotiation

1 Introduction

Internet of things (IoT) and cloud computing are two emerging fields where a lot of research is currently going on. In a cursory glance, these two technologies seem disparate. However, their complementary characteristics lend them well to be integrated into a single system [1]. IoT refers to a worldwide network of interconnected devices which are uniquely addressable and which can gather data about their environment

S. Chatterjee (✉) · S. G. Samaddar
Department of Computer Science and Engineering, National Institute of Technology, Sikkim 737139, India
e-mail: sayanc2011@gmail.com

S. G. Samaddar
e-mail: shefalika99@yahoo.com

© Springer Nature Singapore Pte Ltd. 2020
A. Elçi et al. (eds.), *Smart Computing Paradigms: New Progresses and Challenges*,
Advances in Intelligent Systems and Computing 767,
https://doi.org/10.1007/978-981-13-9680-9_7

101

using sensors and can take actions in the environment using actuators [2]. Thus, common household appliances like fridge, washing machine, etc., can act as individual IoT nodes and be connected through an "Internet of things" with other IoT nodes like mobile phones, self-driving vehicles, etc. As they have limited processing power and battery life, they can benefit immensely by integration into a cloud platform.

Cloud computing is a paradigm under which a client can access networked computing resources elastically. The cloud provider maintains server banks which can be allocated to a cloud client as and when required. The resources in the cloud seem unlimited to the client [3]. IoT nodes, being devices of limited processing power, can leverage unlimited processing power of cloud computing. In fact, many works have already been done on the integration of IoT and cloud [1, 4–6]. Known as Cloud of things, the integrated network of IoT and cloud allows not only the resource-constrained IoT nodes to utilize the processing power of the cloud but also the centralized and fixed cloud to spread into the world of distributed and ubiquitous computing that IoT represents.

Users of the IoT nodes can access personal data remotely from the nodes by using mobile devices like smart phones. However, the data being exchanged between the nodes and the mobile device may be highly personal in nature and hence must travel encrypted. Moreover, the mobile device and the IoT nodes must authenticate each other as well. This requires a low-cost and secure remote authentication and session key generation protocol between the mobile device and the IoT nodes.

This paper presents such a protocol which leverages the power of the cloud in the form of a trusted cloud gateway server which authenticates the IoT nodes and the smart device to each other. The given protocol uses ECC [7–10], which is a low-cost and secure public key cryptographic scheme.

The IoT usage model used in this scheme is shown in Fig. 1. The user first connects to the IoT node using a smart device. After that the IoT node forwards its own and the smart device's credentials to a gateway server present in the cloud. The gateway server authenticates the IoT node and the smart device to each other and helps them generate their session key. As the gateway server is present in the cloud, it can take

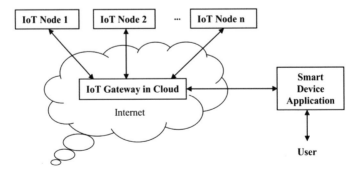

Fig. 1 IoT communication model in cloud

on processing intensive duties of the IoT nodes and the smart device without any problem.

2 Background Study

In the year 1981, Lamport [11] gave an authentication scheme between two parties over an insecure medium. However, certain vulnerabilities were found in his scheme [12]. Since then a lot of work has been done on designing efficient and secure mutual authentication and key generation schemes. In recent years, Das et al. [13] have given a two-factor user authentication method in wireless sensor node environment. Their scheme has used keyed hash and xor function. He et al. [14] have enhanced Das's scheme by improving the scheme's password security. Khemissa et al. [15] have given a keyed hash message authentication protocol between sensor nodes and a base station. Their scheme has achieved mutual authentication with session key generation while having desirable properties like prevention of replay attack, prevention of impersonation attack, etc. However, the base station in their scheme stores a table entry for each node containing the node's id and secret key. Thus, the base station becomes a single point of failure; as, if it is successfully compromised then the whole network becomes compromised. Dhillon et al. [16] have given a mutual authentication and session key generation protocol for wireless sensor network nodes with a gateway server. Though their scheme uses only hash and xor functions, the scheme requires the nodes and the user to have preselected shared secret keys with the gateway server. How the shared secrets are communicated in a safe and secure manner is not elaborated in the scheme. Esfahaniet et al. [17] have devised a mutual authentication and session key generation protocol between a sensor node and smart router using an authentication server in the IoT environment. Their protocol uses only xor and hash functions and provides various security features like forward security, resistance against man-in-the-middle attack, etc. However, their protocol also requires the presence of secure channel to distribute initial secret keys.

3 The Proposed Scheme

The workflow of the proposed scheme is presented through the given UML use case diagram given in Fig. 2.

The user first logs into the smart device using his password and biometrics. The user can also change his login password to the smart device. Once the user logs into the smart device, the smart device connects with an IoT node as instructed by the user. The smart device and all the IoT nodes are registered to a gateway node in cloud. The gateway node does the heavy duty task of authenticating the smart device and the IoT node. At the end of authentication phase, both the smart device and the IoT node have the secret key for that session. All subsequent communications between

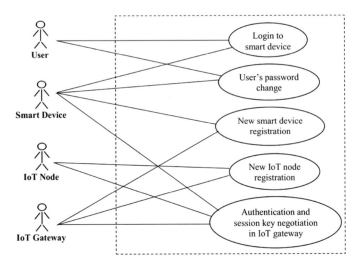

Fig. 2 UML use case diagram for the proposed scheme

the IoT node and the smart device are encrypted by the session key for that session. In the subsequent discussion, the notations shown in Table 1 have been used.

The proposed scheme includes two modules, namely (1) registration phase and (2) mutual authentication and session key negotiation phase, discussed in the following three subsections.

Table 1 Symbols and their meaning

ID_U, ID_D, ID_N	Identity of the user, smart device and IoT node, respectively
G	Generator of the elliptic curve
B_i	Biometric parameter of the user
Pw, Pw^*	User's passwords
$h(\ )$	Cryptographic hash function
$H(\ )$	Perceptual hash function
$\oplus$	xor operation
CA_d, CA_n, CA_g	Public key certificates of the smart device, IoT node and the gateway, respectively
PU_d, PU_n, PU_g	Public keys of the smart device, IoT node and the gateway, respectively
r_d, r_n, r_g	Private keys of the smart device, IoT node and the gateway, respectively
N_i, N_j	Nonces
r, p	Random session secrets of the smart device and the IoT node, respectively
SK	Session key

3.1 System Initialization and Registration Phase

Registration phase of the proposed scheme is further divided into two major modules, namely (1) user's registration with the smart device and (2) registration with the IoT gateway. These are discussed in the following subsections.

User's Registration with the Smart Device User has a smart device through which he/she operates the IoT devices over the Internet. Initially, the user has to register the smart device by providing his/her identity ID_U and biometric parameter B_i and user's secret password P_W. However, the smart device does not store these values in plain text form. Rather, the smart device upon receiving the user's credential calculates the cryptographic hash function of ID_U and P_W and the perceptual hash function of B_i and stores them for user verification at login.

A hash function is a mapping which converts any data of arbitrary length into and a fixed length output. A cryptographic hash function has the additional properties of being irreversible, i.e. given the output, it is infeasible to calculate the input and two inputs varying by even small amount produce outputs which are widely varying from each other so that no correlation between inputs can be divined by looking at hash outputs [7]. SHA-3 is a widely used cryptographic hash protocol [18]. Perceptual hashing differs from cryptographic hashing in that it preserves correlation between inputs so that slightly differing inputs give rise to the same hash output [19]. As biometric readings in different sessions may give rise to slightly differing readings, perceptual hash is most appropriate for biometric parameters.

In addition to the user's registration in the smart device, our scheme also provides *user's login to smart device* and *user's password change* procedures which are described below.

User's Login to Smart Device While logging into the smart device, the user provides his/her identity ID_U, password Pw and biometric reading B_i. The smart device calculates the hash values of the credentials and compares them to the stored hashes. If they match, then the user is logged in.

User's Password Change When the user wants to change his/her password, he enters his identity ID_U, biometric reading B_i, previous password Pw and the new password Pw*. The smart device first calculates and matches the corresponding hash of ID_U, B_i and Pw with the hash values stored in its storage. It then calculates the cryptographic hash value of the new password Pw* and replaces the prior hash of Pw with the hash of Pw*.

Registration with the Gateway Server In this section, we have two submodules, namely (1) smart device registration with the gateway and (2) IoT node registration with the gateway which are discussed below. The given scheme uses elliptic curve cryptography (ECC) as ECC is much more efficient than comparable public key cryptosystem like RSA. An ECC scheme of 160-bit key size gives the same security as 1024-bit key-sized RSA scheme [8]. At the initial stage of the protocol, the gateway server makes public the ECC parameters like the elliptic curve used, its generator G, and the curve's prime order n along with its public key PU_g.

Smart Device Registration with Gateway Server In this phase, the smart device registers itself with the gateway. It first sends its identity and it public key certificate CA_d to the gateway. The gateway stores its ID and public key for future use. In its turn, the gateway sends an acknowledgement and its own public key certificate CA_g to the smart device, which the smart device saves.

IoT Device Registration with Gateway Server The IoT node registers itself with the gateway server following the same procedure as the smart device. The IoT node sends its identity and public key certificate CA_n to the gateway server. The gateway server after verifying the certificate stores the IoT node's ID and public key and sends its own certificate CA_g along with an acknowledgement to the node.

3.2 *Mutual Authentication and Session Key Negotiation*

In this phase, the smart device and the IoT nodes authenticate themselves to each other using the gateway server. The whole procedure is shown in *Fig. 3*. The steps are discussed below:

Step 1 In this step, the smart device generates a nonce N_i. A nonce is a random number which is used only once. The smart device then computes $r_d.PU_g$, where r_d is the private key of the smart device and PU_g is the public key of the gateway. It then computes the hash of the xor of its identity ID_d, the nonce N_i and $r_d.PU_g$ and sends $\{ID_d, N_i, h(ID_d \oplus N_i \oplus r_d.PU_g)\}$ to the IoT node.

Step 2 The IoT node, on receiving the packet from the smart device, generates a random session secret $p \in Z_n^*$. Here Z_n^* is the set $\{1, ..., n-1\}$ where n is the prime order of the elliptic curve. It also computes $r_n.PU_g$, where r_n is its secret key. The IoT node then appends the data $\{ID_n, h(ID_n \oplus r_n.PU_g \oplus N_i), p \oplus h(r_n.PU_g)\}$ with the data it received from *step 1* and sends it to the IoT gateway.

Step 3 The gateway server first computes $r_g.PU_d$, where r_g is its private key and PU_d is the smart device's public key. It then compares $h(ID_d \oplus N_i \oplus r_g.PU_d)$ with the received $h(ID_d \oplus N_i \oplus r_d.PU_g)$. Note that $r_d.PU_g = r_d.r_g.G = r_g.r_d.G = r_g.PU_d$. Hence, if both values are equal, then the gateway can authenticate the smart device.

Now the gateway server computes $r_g.PU_n$ and compares $h(ID_n \oplus r_g.PU_n \oplus N_i)$ to $h(ID_n \oplus r_n.PU_g \oplus N_i)$. As $r_g.PU_n = r_g.r_n.G = r_n.r_g.G = r_n.PU_g$, the equality of the two values prove the authenticity of the IoT node to the gateway.

The server xors $h(r_g.PU_n)$ with the received $p \oplus h(r_n.PU_g)$ to get the value of IoT node's secret p. It generates a nonce N_j and sends $\{N_j, h(r_g.PU_d \oplus N_j), p \oplus r_g.PU_d\}$ to the smart device.

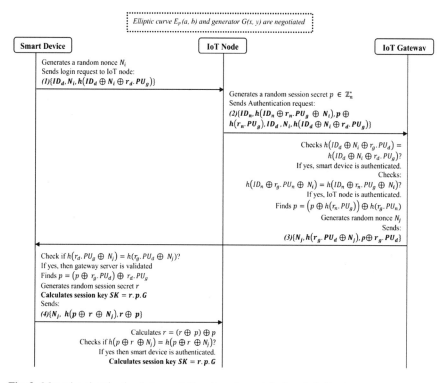

Fig. 3 Mutual authentication between IoT node and smart device via IoT gateway

Step 4 The smart device computes $r_d.PU_g$ and checks whether $h(r_d.PU_g \oplus N_j) = h(r_g.PU_d \oplus N_j)$. As $r_d.PU_g = r_d.r_g.G = r_g.r_d.G = r_g.PU_d$, the equality on both sides authenticates the gateway server to the smart device. The smart device then finds the IoT node's secret p by computing $r_d.PU_g$ and xoring it with the received $(p \oplus r_g.PU_d)$. It then generates a session secret r and calculates the session key $SK = r.p.G$. The smart device then sends $\{N_j, h(p \oplus r \oplus N_j), r \oplus p\}$ to the IoT node.

Step 5 The IoT node now finds the value of r by xoring p with the received $r \oplus p$. It then calculates $h(p \oplus r \oplus N_j)$ and compares it to the received $h(p \oplus r \oplus N_j)$. If both values are equal, then the IoT node becomes sure of the authenticity of the smart device. The smart device can then calculate the session key as $SK = p.r.G$. The session can now commence securely.

4 Security Analysis of the Proposed Scheme

This section gives an informal analysis of the proposed scheme. In discussing the security properties of the given scheme, it is assumed that the channel of communication is insecure, i.e. an intruder can read, capture and modify whatever data is being transmitted through the channel.

4.1 Mutual Authentication

Mutual authentication implies that all the parties involved in the scheme are convinced about the authenticity of each other's identity. The given scheme has this property. The gateway server authenticates the IoT node after receiving *message* 2 and verifying that $h(ID_n \oplus r_g.PU_n \oplus N_i) = h(ID_n \oplus r_n.PU_g \oplus N_i)$. This is possible because r_n is only known to the IoT node. Similarly, the gateway server authenticates the smart device by checking $h(ID_d \oplus N_i \oplus r_g.PU_d) = h(ID_d \oplus N_i \oplus r_d.PU_g)$. This is possible because r_d is only known to the smart device.

The smart device authenticates the gateway server after receiving *message* 3 and checking that $h(r_d.PU_g \oplus N_j) = h(r_g.PU_d \oplus N_j)$. The IoT node authenticates the smart device after receiving *message* 4 and checking if the smart device has correctly computed the value of $h(p \oplus r \oplus N_j)$.

4.2 Confidentiality

In the given scheme, the session key is never sent. The session secrets which form the session key do not travel openly and can only be found by somebody who knows the value of private keys r_d, r_n or r_g. Similarly, any other secret information like private keys, etc., are never sent in the open. Hence, the given scheme maintains confidentiality property.

4.3 Resistance to Man-in-the-Middle Attack

In a man-in-the-middle attack, an intruder captures and alters the data travelling between two communicating parties while the parties believe that they are communicating between themselves. The given scheme is secure against man-in-the-middle attack as altered data can be detected by the communicating parties.

For example, in *message* 2 if an intruder replaces IoT node's masked session secret $p \oplus h(r_n.PU_g)$ with some other value, then after receiving *message* 4, the IoT node can detect the change as $h(p \oplus r \oplus N_j)$ will not match the received value. This is also true if an intruder changes the value of r in $r \oplus \boldsymbol{p}$ in *message* 4.

4.4 Resistance to Replay Attack

In a replay attack, an intruder gets hold of a message and resends it at a later time. The use of nonce in the given scheme prevents the use of replay attack. Nonce is a random number used only once. For example, the use of nonce N_i within a hash function in *message* 1 and 2 means that the gateway server can detect if the message is replayed at a later time. Similarly, in *message* 3 the gateway server sends a nonce N_j and includes it in the hash function so that the message cannot be retransmitted to the smart device or the IoT node at a later time.

4.5 Perfect Forward Security

A protocol is said to have perfect forward security if even after the long-term keys become known at certain point of time, the sessions before that point of time still remain undecipherable. The given scheme has perfect forward security as the session keys for different sessions depend on different random variables p and r.

4.6 Resistance Against Known Session Key Attack

A protocol is vulnerable to known session key attack if knowing the session key of a particular session allows an intruder to guess the session keys of other sessions. This is not possible in the given scheme as the session keys are independent of each other.

5 Computation Cost Analysis

This section gives computation cost of our scheme. The given scheme uses very low-cost hash operations and ECC point multiplication. It is well known that due to small key size, ECC operations are much cheaper than corresponding RSA operations [8].

Let the time to compute one biometric perceptual hash operation, one normal hash operation, one point multiplication operation, one RSA exponentiation operation, one xor operation and one symmetric encryption/decryption operation be represented by T_H, T_h, T_{PM}, T_{EX}, T_{XOR}, $T_{E/D}$ respectively. Then $T_{EX} \gg T_{E/D} \gg T_{PM} > T_H > T_h$

Table 2 Number of operations done by the smart device, IoT node and gateway in different phases of the protocol

Phases devices	User registration with smart device	User login with smart device	User password change	Mutual authentication and session key negotiation
Smart device	$T_H + T_h$	$T_H + T_h$	$T_H + 2T_h$	$3T_h + 3T_{PM}$
IoT node	–	–	–	$3T_h + 3T_{PM}$
IoT gateway	–	–	–	$4T_h + 2T_{PM}$

$> T_{XOR}$. Since time to compute one xor operation is minute in comparison with the other operations, it is ignored in the following analysis.

Table 2 gives number of operations done by the IoT node, the smart device and the gateway server in the different stages of the protocol. The given protocol manages to correctly authenticate and securely generate session keys while consuming very little computational resource.

6 Conclusion

The protocol given in this paper is a very lightweight solution to the problem of three-way mutual authentication and session key generation among resource-constrained clients. As such, though the protocol has been given for authentication between an IoT node, smart device and cloud gateway server, it has broad applicability in other scenarios as well. The given protocol does not use any encryption/decryption operations or other difficult to implement features like timestamp. However, the protocol satisfies most of the pertinent security properties expected from such a scheme.

References

1. A. Botta, W. De Donato, V. Persico, A. Pescapé, Integration of cloud computing and internet of things: a survey. Future Gener. Comput. Syst. **56**, 684–700 (2016)
2. R.H. Weber, R. Weber, in *Internet of Things,* vol. 12 (Springer, New York, NY, USA) (2010)
3. P. Mell, T. Grance, in *The NIST Definition of Cloud Computing* (2011)
4. T. Bhattasali, R. Chaki, N. Chaki, in *Secure and Trusted Cloud of Things.* Annual IEEE In India Conference (INDICON) (2013), pp. 1–6
5. J. Zhou, T. Leppanen, E. Harjula, M. Ylianttila, T. Ojala, C. Yu, L.T. Yang, in *CloudOfThings: A Common Architecture for Integrating the Internet of Things with Cloud Computing.* IEEE 17th International Conference Computer Supported Cooperative Work in Design (CSCWD) (2013), pp. 651–657
6. B. Kantarci, H.T. Mouftah, Trustworthy sensing for public safety in cloud-centric internet of things. IEEE Internet Things J. **1**(4), 360–368 (2014)

7. C. Paar, J. Pelzl, in *Understanding Cryptography: A Textbook for Students and Practitioners* (Springer Science & Business Media, 2009)
8. D. Hankerson, A.J. Menezes, S. Vanstone, in *Guide to Elliptic Curve Cryptography* (Springer Science and Business Media, 2006)
9. V.S. Miller, in *Use of Elliptic Curves in Cryptography*. Conference on the Theory and Application of Cryptographic Techniques (Springer, Berlin, Heidelberg, 1985), pp. 417–426
10. N. Koblitz, Elliptic curve cryptosystems. Math. Comput. **48**(177), 203–209 (1987)
11. L. Lamport, Password authentication with insecure communication. Commun. ACM **24**(11), 770–772 (1981)
12. C.J. Mitchell, L. Chen, Comments on the S/KEY user authentication scheme. ACM SIGOPS Operating Syst. Rev. **30**(4), 12–16 (1996)
13. M.L. Das, Two-factor user authentication in wireless sensor networks. IEEE Trans. Wireless Commun. **8**(3), 1086–1090 (2009)
14. D. He, Y. Gao, S. Chan, C. Chen, J. Bu, An enhanced two-factor user authentication scheme in wireless sensor networks. Ad hoc Sens. Wireless Netw. **10**(4), 361–371 (2010)
15. H. Khemissa, D. Tandjaoui, in *A Novel Lightweight Authentication Scheme for Heterogeneous Wireless Sensor Networks in the Context of Internet of Things*. Wireless Telecommunications Symposium (WTS) (IEEE, 2016), pp. 1–6
16. P.K. Dhillon, S. Kalra, A lightweight biometrics based remote user authentication scheme for IoT services. J. Inf. Secur. Appl. (2017)
17. A. Esfahani, G. Mantas, R. Matischek, F. B. Saghezchi, J. Rodriguez, A. Bicaku, J. Bastos, A lightweight authentication mechanism for M2M communications in industrial IoT environment. IEEE Internet Things J. (2017)
18. M.J. Dworkin, SHA-3 standard: Permutation-based hash and extendable-output functions. Fed. Inf. Process. Stds. (NIST FIPS)-202 (2015)
19. R. Venkatesan, S.M. Koon, M.H. Jakubowski, P. Moulin, in *Robust Image Hashing*. Proceedings of IEEE International Conference Image Processing, vol. 3 (2000), pp. 664–666

Better Quality Classifiers for Social Media Content: Crowdsourcing with Decision Trees

Ian McCulloh, Rachel Cohen and Richard Takacs

Abstract As social media use grows and increasingly becomes a forum for social debate in politics, social issues, sports, and brand sentiment; accurately classifying social media sentiment remains an important computational challenge. Social media posts present numerous challenges for text classification. This paper presents an approach to introduce guided decision trees into the design of a crowdsourcing platform to extract additional data features, reduce task cognitive complexity, and improve the quality of the resulting labeled text corpus. We compare the quality of the proposed approach with off-the-shelf sentiment classifiers and a crowdsourced solution without a decision tree using a tweet sample from the social media firestorm #CancelColbert. We find that the proposed crowdsource with decision tree approach results in a training corpus with higher quality, necessary for effective classification of social media content.

Keywords Social media · Sentiment · Classifier · Machine learning · Decision tree · Twitter · Turk

1 Introduction

Developing automated classifiers for social media content is an important problem for data scientists and privacy researchers. People are increasingly using social media to express opinions on a wide range of issues spanning politics, social injustice, corporations, sports teams, and more. User's opinions can vary, ranging from support to opposition. Data scientists may apply automated classifiers to these data to understand public sentiment toward certain brands or social issues that are being debated

I. McCulloh (✉) · R. Cohen · R. Takacs
Johns Hopkins University, 11100 Johns Hopkins Road, 20723 Laurel, MD, USA
e-mail: ian.mcculloh@jhuapl.edu

R. Cohen
e-mail: rachel.cohen@jhuapl.edu

R. Takacs
e-mail: richard.takacs@jhuapl.edu

© Springer Nature Singapore Pte Ltd. 2020
A. Elçi et al. (eds.), *Smart Computing Paradigms: New Progresses and Challenges*,
Advances in Intelligent Systems and Computing 767,
https://doi.org/10.1007/978-981-13-9680-9_8

online. Privacy researchers may be interested in using these classifiers to identify specific individuals within those online discussions to target key users or communities for influence interventions. Developing automated classifiers to measure support and opposition is complicated by several factors to include short size of text [1–3], sarcasm [3–7], humor [5, 7, 8], political alignment [9], emoticons [10], among other factors. We investigate methods to overcome these issues using crowdsourcing and decision trees.

Twitter is an online news and social networking service that allows users to post and interact with 280-character messages called "tweets". As of the second quarter of 2017, the microblogging service averaged 328 million monthly active users, making it the eighth most popular social network in the world [11]. As the number of users continues to rise, Twitter, as a social platform, demonstrates the ability of social media to influence politics and social debate. As a result, Twitter's rich data can provide insight as to how the general public perceives a topic. With that value realized, a single tweet object can serve as a trove of metadata, including the tweet's content, the location it was sent from, and the time it was sent. Twitter provides a well-documented application programming interface (API) as a way to query Twitter data and fetch results in a standardized format, which can then be parsed for areas of specific interest.

Developing text classifiers for Twitter data using a supervised learning approach requires a gold-standard training dataset to develop and evaluate the veracity of potential classifiers. While there exist several off-the-shelf text classifiers for Twitter data [15–19], they may not be tailored to specific applications. We posit that the nature of an online firestorm (large, negative, online discourse) may be fundamentally different than the nature of discourse comprising the off-the-shelf training corpus. Furthermore, the performance of various off-the-shelf classifiers may differ and exhibit varied performance when applied to newly captured data. Finally, we posit that achieving agreement among humans classifying tweets in emotionally charged firestorms is highly problematic due to personal bias and conflation of sentiment, position, humor, sarcasm, and other challenges of assessing micro-blog data.

In this paper, we propose a method to develop a high-quality, gold-standard training corpus tailored for firestorms. The proposed method utilizes crowdsourcing and guided decision trees to aid people in systematically labeling tweets. We contrast this approach with off-the-shelf classifiers and crowdsourcing without decision trees. We demonstrate that crowdsourcing with guided decision trees improves the quality and feature space of the training corpus for developing automated classifiers.

2 Background

This work contributes to the study of online firestorms. Firestorms are defined as "an event where a person, group, or institution suddenly receives a large amount of negative attention [online]" [12]. A firestorm can be characterized by an instance where sudden negative attention is in response to a recent action and arises without prior discussion. For the purpose of this research, the focus was placed on online protest and social debate—these events are often fast moving and often have part in influencing the public's perception of an issue.

This paper will focus on a specific firestorm that erupted in 2014. The firestorm surrounds comedian, Stephen Colbert, and his show *The Colbert Report* on Comedy Central. The hashtag #CancelColbert was given to the controversy that began after a tweet was sent from a Twitter account associated with Colbert's then show. The tweet, that was offensive to many Asian-Americans, seemed to many as a mockery of Asian speech. A certain individual, activist Suey Park, started the campaign with a single tweet: "The Ching-Chong Ding-Dong Foundation for Sensitivity to Orientals has decided to call for #CancelColbert. Trend it" [13]. The hashtag grew in popularity and ultimately made it on Twitter's list of trending topics for a nontrivial period, but it did not come without response from supporters of Colbert in defense of his comedic style. Because of this polarity, this firestorm shows the importance of detecting sentiment around a given hashtag, as individuals can tweet (using the hashtag) having strayed from the sentiment that the original author had hoped for.

The data used in this paper are selected from a corpus of 80 firestorms presented by Lambda et al. [12]. Their data were obtained using Twitter's decahose, which represents an approximately 12% random sample of the Twitter content associated with the #CancelColbert firestorm. The additional 2% above the 10% of tweets is obtained by extracting retweet and mention messages from the 10% random sample of the corpus. Their #CancelColbert firestorm sample consisted of 10.1 MB of data and included 15,591 unique tweets. A sample of 200 tweets from this corpus was selected at random for use in an Amazon Mechanical Turk (AMT) experiment and in comparison with off-the-shelf classifier performance. While limitations of Twitter's sampling methodology are noted [14], these data represent a sufficient corpus for the purpose of evaluating construction of a gold-standard training set.

Several off-the-shelf classifiers exist for assessing sentiment within Twitter data. AFINN, from Finn Årup Nielsen, is an English wordlist-based approach for sentiment analysis. The AFINN lexicon assigns words with a score that runs between -5 and 5, with negative scores indicating negative sentiment and positive scores indicating positive sentiment [15, 16]. The NRC Emotion Lexicon, from Saif Mohammad and Peter Turney, is a list of English words and their associations with eight basic emotions (anger, fear, anticipation, trust, surprise, sadness, joy, and disgust) and two sentiments (negative and positive). This lexicon categorizes words in a binary fashion ("yes", "no") if it fits into one of the emotion categories [17, 18]. The Bing classifier, by Liu et al., categorizes words in a binary fashion ("positive", "negative") [19]. When applied to a data corpus, however, each of the off-the-shelf sentiment classifiers differs somewhat in their assessment of sentiment. There are many potential reasons for this such as how classifiers treat the presence of sarcasm, humor, and colloquial symbols or text. Differences in classifier performance, however, bring into question the veracity of a given classifier for firestorm sentiment analysis. Developing a tailored classifier for firestorms requires the construction of a gold-standard training corpus.

AMT is an on-demand crowdsourcing marketplace that allows individuals to request work from others online. This marketplace allows people to complete "human intelligence tasks" (HIT)—tasks that humans can currently do more intelligently than computers. It allows requesters to crowdsource data from tasks ranging from object detection in photos to text translation. The requester of the work specifies how many workers can complete a task, determines a monetary value to reward them with, and for the case of sentiment analysis of tweets, asking several workers to provide annotations for each tweet will improve the accuracy of the results. By having multiple annotators assess the tweets provided in the task, there will be random overlap of workers annotating the same tweet.

An important measure of quality in training data is the inter-annotator agreement (IAA). IAA is the level of agreement between raters (annotators), which is high if all raters consistently agree when independently labeling data and low when they disagree. Krippendorff's Alpha is best suited for this data because it can be adjusted for a variable number of annotators assessing different tweets, handles missing data, and is uniformly more powerful than competing methods [20–23]. Missing data will be an important consideration when using a decision tree approach, where annotators have the option of choosing *"Not Applicable"* when coding text labels. Alpha (α) is given by:

$$\alpha = 1 - \frac{D_o}{D_e}$$

where D_o is the disagreement observed:

$$D_o = \frac{1}{n} \sum_{c \in R} \sum_{k \in R} \delta(c, k) \sum_{u \in U} m_u \frac{n_{cku}}{P(m_u, 2)}$$

and D_e is the expected disagreement:

$$D_e = \frac{1}{P(n, 2)} \sum_{c \in R} \sum_{k \in R} \delta(c, k) P_{ck}$$

where P_{ck} is the number of possible pairs that could be made. Here, $\alpha = 1$ indicates 100% reliability and $\alpha = 0$ indicates a complete lack of reliability [22].

The resulting α ranges from -1.0 to 1.0, where a 1.0 indicates perfect agreement and -1.0 indicates perfect disagreement. The benchmark for random chance agreement scales with the number of annotators and converges to 0.0 as the number of annotators becomes large. Beyerl et al. [21] proposed a benchmark of quality using Krippendorff's alpha, where $0.66 < \alpha < 0.80$ is good agreement, corresponding to approximately 80% probability of joint agreement (PJA) and $\alpha > 0.80$ is excellent agreement, corresponding to better than 90% PJA.

3 Methods

Research consisted of three phases. The first phase involved exploring the competing performance of the three off-the-shelf sentiment classifiers. The second phase involved conducting an AMT experiment to compare IAA with the off-the-shelf classifiers. The third phase involved incorporating a guided decision tree within Amazon Mechanical Turk to extract additional data features and structure the responses of the crowdsourced annotators. For each phase of analysis, we used the random sample of 200 #CancelColbert tweets from the Lambda et al. [12] firestorm corpus. The same dataset is used across all phases for more accurate comparison and was limited to a sample of 200 for cost efficiency. IAA is calculated using Krippendorf's Alpha for sentiment ratings of positive/negative/neutral across the four off-the-shelf sentiment classifiers. IAA is calculated for the second phase of the Amazon Mechanical Turk experiment positive/negative/neutral and benchmarked against off-the-shelf classifiers. Finally, IAA is calculated for the third phase of the AMT experiment using guided decision trees.

AMT was used to crowdsource coding of tweets for phases two and three. AMT annotators were paid a reward of $0.07 per HIT. Each tweet was coded by nine independent annotators. Annotators coded between one and 20 tweets. All annotators were "Masters", who are seen as top workers in the AMT marketplace and have been awarded this qualification due to their high degree of success across different AMT tasks.

For phase two, annotators were asked to code tweets on a five-point Likert scale, ranging from strongly positive to strongly negative. They were also provided the option of "I Don't Know". For each option, the AMT worker was provided guidance for their choice. For example, the strongly positive rating was accompanied by the guidance, "Select this if the item embodies emotion that contains extreme support, satisfaction, or general happiness. For example, 'This candidate is going to change our country for the better. I support them all the way'". In the authors' prior work using AMT, providing workers guidance increased the inter-annotator agreement when viewing responses. This guidance is provided to establish the best-case IAA score in comparison with the guided decision tree approach.

For the guided decision tree, annotators were asked to respond to four questions. For two of the four questions, the annotator was asked to respond to a sub-question depending upon their response. The questions for the guided decision tree were:

1. Does the author of the tweet express a certain position regarding canceling The Colbert Show?

    a. If yes, what is the author's position regarding canceling The Colbert Show?

2. Does the author of the tweet convey a clear sentiment toward the subject of the tweet?

    a. If yes, is the sentiment positive or negative?

    b. If no, is the tweet a straightforward reporting of facts or a headline with/without a link?

3. Did the author of the tweet use sarcasm to mock or convey contempt toward the subject of the tweet?
4. Did the author of the tweet use humor to provoke laughter or amusement?

Prior to beginning the AMT task, subjects acknowledged the following instruction:

The purpose of this project is to analyze the sentiment of politically, socially, and/or culturally charged tweets. Please try and identify them in this context. Keep in mind that you are to rate the Twitter user's sentiment or position, and that you are NOT rating the tweet based on your personal feelings on the subject. Please answer the following questions for a given tweet.

Structuring the AMT crowdsourcing task in this manner allows additional features to be captured for the training corpus and is likely to screen out ambiguous sentiment in the cognition of the annotator. We posit that this will improve IAA.

Ethics approval for this study was approved by the Johns Hopkins University Institutional Review Board (IRB) for use of AMT workers in the crowdsourcing tasks. All AMT workers were presented with the background and purpose of the study and could opt-out at any time. AMT workers acknowledged informed consent prior to participation in the project.

4 Findings

The IAA rating for off-the-shelf sentiment classifiers was $\alpha = 0.525$. This rating falls below acceptable limits according to the Bayerl scale. The IAA rating for the basic AMT experiment was $\alpha = 0.301$. The rating could be improved slightly to $\alpha = 0.308$ by including the "I Don't Know" as a response choice and excluding ratings that were completed in 3 s or less. It is surprising that the IAA rate for the AMT experiment performs worse than the off-the-shelf sentiment classifiers. Consistency among off-the-shelf classifiers with each other, however, does not mean that those classifiers are more accurate at classifying sentiment within this specific Twitter firestorm. It could represent a systematic bias. In any case, the best-case scenario still falls below established quality standards for use.

The results of the IAA for the guided decision tree are displayed in Fig. 1. It can be seen that all questions meet the Bayerl scale for good agreement and questions 1A, 2B, and 4 exceed the benchmark for excellent agreement. These questions require AMT workers to accurately assess tweet position, lack of sentiment, and humor, given that they accurately recognized whether a tweet expressed a position, sentiment, or humor. The guided decision tree allows the data scientist to effectively control for recognition when assessing IAA on coding/label assignment to Twitter data.

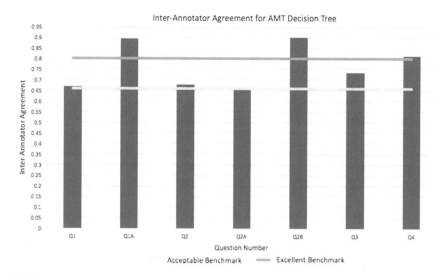

Fig. 1 Inter-annotator agreement (IAA) for guided decision tree crowdsourcing approach. This shows that all IAA ratings exceed the acceptable benchmark of 0.66 and several exceed excellent benchmark of 0.80

5 Conclusion

Constructing a Twitter training corpus using crowdsourcing with a guided decision tree appears to improve the quality of the corpus over off-the-shelf or simple crowd-sourcing approaches. The decision tree effectively broke down a complex task into smaller problems that allowed AMT workers to achieve higher rates of agreement. It also provided a means for raters to remove bias when evaluating different types of tweets by first cognitively assessing the position and tone of the tweet prior to rendering a judgment. The ability to detect a positive, negative, or neutral sentiment remains a nuanced process, however, especially when language constructs such as humor and sarcasm are in play. Since these language paradigms are often difficult for a machine or artificial intelligence to detect, it is notable that AMT workers had a higher percentage of agreement when rating tweets for these factors in the guided decision tree approach. Arguably, linguistic factors such as humor and sarcasm are just as important (if not more important) as detecting sentiment, especially in a firestorm context surrounding a social debate. It is also notable that when AMT workers were not guided through a decision tree, the presence of humor and sarcasm degraded their IAA rating.

This work faced several limitations. The AMT experiment was focused on a single firestorm corpus and utilized a random sample of 200 tweets. Future research may investigate additional firestorm corpora and may include random samples with more tweets. The required sample size estimation could be performed by investigating the sensitivity of IAA rating by randomly removing sampled tweets. Similarly, the same decision tree structure could be applied to different corpora to evaluate the sensitivity of the data corpus on the findings.

Despite these limitations, this paper further demonstrates the inherent limitations of off-the-shelf classifiers. It highlights unique challenges for sentiment classification present within online protests. As Internet activism flourishes due to the immediacy of social media, the hasty spread of information, and political engagement; construction of high-quality training corpora becomes more important. These data challenges call for more work in classifying additional data features such as emotion, sarcasm, humor, polarizing positions, and even "hijacked hashtags". The crowdsourced, decision tree approach presented in this paper has proven effective in developing a high-quality gold-standard training dataset that outperforms off-the-shelf solutions.

Acknowledgements This work was supported by the Office of Naval Research, Grant No. N00014-17-1-2981/127025

References

1. E. Kouloumpis, T. Wilson, J.D. Moore, Twitter sentiment analysis: The good the bad and the omg!, in *ICWSM 11* (2011), pp. 11538–541
2. A. Bermingham, A.F. Smeaton, Classifying sentiment in microblogs: Is brevity an advantage?, in *19th ACM International Conference on Information and Knowledge Management, ACM* (2010), pp. 1833–1836
3. H. Yu, V. Hatzivassiloglou, Towards answering opinion questions: Separating facts from opinions and identifying the polarity of opinion sentences, in *EMNLP* (2003)
4. D. Maynard, M.A. Greenwood, Who cares about sarcastic tweets? Investigating the impact of sarcasm on sentiment analysis, in *Language Resources and Evaluation Conference* (2014), pp. 4238–4243
5. A. Reyes, P. Rosso, T. Veale, A multidimensional approach for detecting irony in Twitter, in *Language Resources and Evaluation*, vol. 47, no. 1 (2013), pp. 239–268
6. O. Tsur, D. Davidov, A. Rappoport, A great catchy name: Semi-supervised recognition of sarcastic sentences in online product reviews, in *Fourth International AAAI Conference on Weblogs and Social Media* (2010)
7. R.W. Gibbs, H.L. Colston, in *Irony in Language and Thought* (Routledge (Taylor and Francis), New York, 2007)
8. C. Bosco, V. Patti, A. Bolioli, Developing corpora for sentiment analysis: The case of irony and senti-TUT. IEEE Intell. Syst. **28**(2), 55–63 (2013)
9. S. Park, M. Ko, J. Kim, Y. Liu, J. Song, The politics of comments: Predicting political orientation of news stories with commenters' sentiment patterns, in *ACM 2011 Conference on Computer Supported Cooperative Work, ACM* (2011), pp. 113–122
10. K.-L. Liu, W.-J. Li, M. Guo, Emoticon smoothed language models for twitter sentiment analysis, in *AAAI* (2012)
11. J. Dunn, Facebook totally dominates the list of most popular social media apps, [online] Business Insider, Available at http://www.businessinsider.com/facebook-dominates-most-popular-social-media-apps-chart-2017-7 (2017)
12. H. Lamba, M.M. Malik, J. Pfeffer, A tempest in a teacup? Analyzing firestorms on twitter, in *Advances in Social Networks Analysis and Mining (ASONAM), 2015 IEEE/ACM International Conference, IEEE* (2015), pp. 17–24

13. J. Kang, Campaign to 'Cancel' Colbert, [online] The New Yorker, Available at https://www.newyorker.com/news/news-desk/the-campaign-to-cancel-colbert (2014)
14. F. Morstatter, J. Pfeffer, H. Liu, When is it biased?: Assessing the representativeness of twitter's streaming API, in *23rd International Conference on World Wide Web ACM* (2014), pp. 555–556
15. M.M. Bradley, P.J. Lang, Affective norms for english words (ANEW) instruction manual and affective ratings, Technical Report C-1, The Center for Research in Psychophysiology University of Florida (2009)
16. F.Å. Nielsen, A new ANEW: Evaluation of a word list for sentiment analysis in microblogs, arXiv preprint arXiv:1103.2903 (2011)
17. X. Zhu, S. Kiritchenko, S. Mohammad, NRC-Canada-2014: Recent improvements in the sentiment analysis of tweets, in *SemEval@ COLING* (2014), pp. 443–447
18. S. Kiritchenko, X. Zhu, S.M. Mohammad, Sentiment analysis of short informal texts. J. Artif. Intell. Res. **50**, 723–762 (2014)
19. B. Liu, Sentiment analysis and opinion mining, in *Synthesis Lectures on Human Language Technologies*, vol. 5, no. 1 (2012), pp. 1–167
20. R. Artstein, M. Poesio, Inter-coder agreement for computational linguistics. Comput. Linguist. **34**(4), 555–596 (2008)
21. P.S. Bayerl, K.I. Paul, What determines inter-coder agreement in manual annotations? A meta-analytic investigation. Comput. Linguist. **37**(4), 699–725 (2011)
22. A.F. Hayes, K. Krippendorff, Answering the call for a standard reliability measure for coding data, in *Communication Methods and Measures*, vol. 1, no. 1, pp. 77–89 (2007)
23. K.A. Neuendorf, *The Content Analysis Guidebook* (Sage, 2016)

Emerging Techniques in Computing

Predicting Manufacturing Feasibility Using Context Analysis

Vivek Kumar, Dilip K. Sharma and Vinay K. Mishra

Abstract In the present scenario, the technology revisions are making the market vulnerable to predict. This gives birth to the requirement of a new model which can stop loss of a production business by predicting the market using news analysis, associative rule mining, precise predicting techniques and context analysis. This paper presents a novel idea of dealing with manufacturers' problem of product dump due to the rapid change in technology and the changing demand of customers. Every new product launched with new features or with existing features but less price gives a tough competition to already existing products in the market. By the time the manufacturer comes to know that the demand has been decreased, the manufacturer is already in the loss and he has to dump already manufactured pieces due to rapid down sale. In this paper, a model is proposed with an algorithm to quickly identify the required number of pieces in a time frame.

Keywords Analytics · Associative rule mining · Context mining · Predictive analytics · Manufacturing · Product dump

1 Introduction

Manufacturers nowadays are facing a problem of product dump due to rapid change in technology. They are facing competition with every new product, launching in the market, in a day or week, with new features in the same price or same features in less price. Especially if we speak of mobile and electronics market, this is the large problem for manufacturers.

V. Kumar (✉) · D. K. Sharma
GLA University, 17 Km Stone, NH-2, Mathura-Delhi Road, 281406 Mathura, UP, India
e-mail: vivek.kumar@gla.ac.in

D. K. Sharma
e-mail: dilip.sharma@gla.ac.in

V. K. Mishra
SRMGPC, Tiwariganj, NH-28, Faizabad Road, 227105 Lucknow, UP, India
e-mail: mishravinay78@gmail.com

© Springer Nature Singapore Pte Ltd. 2020
A. Elçi et al. (eds.), *Smart Computing Paradigms: New Progresses and Challenges*,
Advances in Intelligent Systems and Computing 767,
https://doi.org/10.1007/978-981-13-9680-9_9

1.1 Context Analysis

Context analysis is the analysis of correlated attributes that affect each other. In context analysis, we have plenty of data for analysis but during the analysis of real-time data, the analyzer finds difficulty in extracting the context. Below given are few questions for you to find the context:

1. Black and shiny
2. Facing cool breeze

Your answers will be:

1. Hair
2. A place surrounded by snow

Why your mind not answered:

1. Skin or something else
2. Refrigerator or Air Conditioner

This is basically the feeding that mind learns from learning, the medium may be television, listening, culture, surroundings, social media, general knowledge, environment, etc. The question is how do we find the context in daily life? The below-given examples give the better idea of its applications:

Manipulating geometric (spatial) objects in the computer is a challenge in many disciplines like cartography, geology, computer-aided design (CAD) and others [1]. Many contexts are derived from graph analysis related to social media and spatial analysis related to geography.

Porter defines competitive strategy as business thinking which provides a strong foundation for developing corporate strategy. The competitive strategy offers rational and straightforward methods of differentiation, overall cost leadership and focus [2]. These three methods prevent a company from the situation of confusion, among which differentiation is the one which is addressed in this paper and it is recommended to use the below stated system for finding context.

1.2 Predictive Analytics

Prediction is something which fascinates everybody if we talk in the context of life prediction. On the same side, the prediction is the study of present and past with some rules associated which can predict future events containing threats and opportunities. "Give me a long enough lever and a place to stand and I can move the Earth." said by Archimedes in 250 B.C., which proves that if we can make some protocols out of the past studies which can work as a lever then we can also accomplish a big task like turning a business in the direction which will be equal to moving earth.

Predictive Analytics is the technique practiced in just a few places to a competitive weapon in businesses. It was always a part of the business in one way or other to identify the requirements, interest of consumers and purchasing power with the technology changing with the decade. One should have enough data to predict the market interest and trends. Change in technology and lifestyle makes it harder to predict the change in fashion and various others. Big data making the prediction easier because we always have enough data or sometimes more than enough to analyze that what was happening in the last few years or months or days and to predict that what is going to happen in coming days. The more data we have, the more precisely we can predict. Big data is an ever-improving tool for data analysis.

There are various factors which may affect prediction, listed but not limited to, like lifestyle, government policies, foreign policies, environmental change, daily life incidents, etc. To understand the effect, consider the given example:

A mobile phone company, say X, has to predict the number of a mobile phone model, say M-348, to be manufactured. The analysis done on empirical data predicts that 10,000 pieces will be consumed in the market and the company starts manufacturing 3G compatible mobiles. In the meanwhile, the government launches a new policy in association with another company, say Y, saying that the government is providing free Internet for three months for 4G compatible mobile phones. What will happen to the production of M-348? The answer is that the demand of M-348 will die. Out of 10,000 pieces, major part will get dumped in a warehouse due to the rapid change of requirement of 4G compatible handsets. 3G compatible handsets will be marked as outdated by consumers and in turn by market too.

The prediction does not work if we do not add sufficient conditions like government policies, market trends, surveys, reviews and users' interest in a particular product, etc. The way predictive model produce value is simple in concept as they make it possible to make more right decisions, more quickly, and with less expense. But the precision and accuracy of a prediction depend on the size of data, the factors included and the rules associated. The decisions made are then more accurate, efficient and effective. The decisions may be made by humans with the help of data visualization tools or machines by automating the entire decision-making process.

2 Literature Survey and Research Gap

The pervasive computing, as described by the author, is a computing which is people-oriented and context-aware; association rule mining takes only data into consideration and ignores context-aware information [3]. Processes in manufacturing are characterized by high frequency of changes [4], as the author talks about documentation complexity and says that adding context awareness to unstructured event logs improves the results. It has been observed that there is a rapid growth in mining manufacturing processes and enterprises in the last three years [5]. In telecommunication, association rules can be used to predict the failure by identifying root causing events occurring just before failure [6]. In [7], the author developed a rule generation algorithm to extract knowledge in the form of rule. This knowledge presents association rules in the form of IF-THEN rules. Authors in [8], applied data mining to customer response data for its use in design in product families. In [9], the author speaks about feature relation association and object form association as an application of the proposed intelligent design retrieval system. The author in [10] extended their work and applied association rule mining technique to deal with product and process variety mapping. The association rules then designed were deployed to the support production planning of product families within existing production processes. In [11], the author focuses on continually changing demands of customers and associated supply chain management of an enterprise. The author in [12] emphasizes on productions as the most popular means of knowledge representation in Information System and also suggests a variant of implementation of analytic subsystem infrastructure for embedding applications. In [13], the author studies the case of manufacturing enterprise producing and selling coffee machines and proposes five mining algorithms. These algorithms are compared on fitness, precision, generalization and simplicity. It has been found that there is a little work done on identifying the number of units to be manufactured in a time frame. In the next section, it is proposed that news analysis in real time and association with empirical data may give better results.

3 Proposed Model

The model proposed here predicts on the basis of the association of objects related to each other in some context. In this model, industry trends are analyzed for any new competitor or product entering the market. Figure 1 illustrates that how data accumulation from various sources and its analysis, prediction and association rule mining helps manufacturing firms to stop loss at an early stage. The product launch news will be extracted and analyzed in real time and stored in the context database

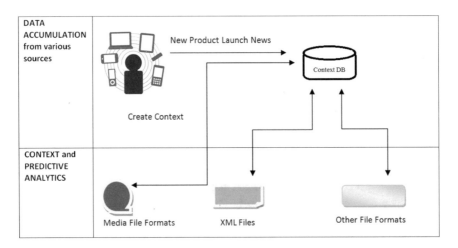

Fig. 1 Model for context analysis to stop loss in manufacturing

for real-time analysis and future analysis as well. Empirical data already stored in XML, Media or other formats, will be combined with the context database.

The data accumulation layer provides news from multiple sources and stores it in the context database for analysis purpose. Context analysis is done as per the algorithm stated in the next section and then decisions are produced as output. Prediction layer is used for predicting the manufacturing speed.

4   Algorithm

The following algorithm may be used to find context and make whatever predictions are possible:

1. The news streaming directly, related to new product launch, is stored in a database for analysis.
2. It is then combined with media news to find the context.
3. If context matches:

   a. Deep digging of features and prices is done from some other files or database containing detailed information of product category or product.
   b. The market trend is analyzed and percentage of effect is calculated.
   c. The predictive analytics is conducted, by assuming that the product is launched today, the number of days is calculated for the sale that when the sale will go down and by what percentage.
   d. If the relative trend is negatively correlated:
      i.   The production unit is ordered to either stop manufacturing or scale down.
      ii. The manual feasibility study is initiated as fresh market opportunities.
      Else:
      iii. The production unit is ordered to scale up or maintain the same pace of manufacturing process.
      iv. The manual feasibility study is initiated as fresh market opportunities.

4. Exit.

5   Conclusion

Associative rule mining and context analysis, when combined with predictive analytics, becomes a powerful tool for any industry or organization. This paper is a hope for manufacturing industries that may save their billions of dollars if they can analyze trends and requirements of the market at the right time. The analysis may stop loss of production units at an early stage like "Stop Loss" feature of stock markets. There are many other areas, like academics, agriculture, etc., on which this model may be applied.

References

1. C. Gold, *Spatial Context: An Introduction to Fundamental Computer Algorithms for Spatial Analysis* (CRC Press, Croydon, 2016)
2. M.E. Porter, *Competitive Strategy: Techniques for Analyzing Industries and Competitors* (The Free Press, New York, 2011)
3. W. Yang, Q. Liao, C. Zhang, An association rules mining algorithm on context-factors and users' preference, in *2013 5th International Conference on Intelligent Human-Machine Systems and Cybernetics*, Hangzhou, China, 2013
4. T. Becker, W. Intoyoad, Context aware process mining in logistics, in *Procedia CIRP*, vol. 63, Budapest, Hungary, 2017, pp. 557–562
5. A.K. Choudhary, M.K. Tiwari, J.A. Harding, Data mining in manufacturing: A review based on the kind of knowledge. J. Intell. Manuf. 501–521
6. J. Hipp, U. Guntzer, G. Nakhaeizadeh, Algorithms for association rule mining: A general survey and comparison, ACM SIG KDD Explorations **2**(1), 58–64
7. J. Buddhakulsomsiri, Y. Siradeghyan, A. Zakarian, X. Li, Association rule generation algorithm for mining automotive warranty data. Int. J. Prod. Res. **44**(14), 2749–2770 (2006)
8. B. Agard, A. Kusiak, Data-mining based methodology for the design of product families. Int. J. Prod. Res. **42**(15), 2955–2969 (2004)
9. C.Y. Tsai, C.A. Chang, Fuzzy neural network for intelligent design retrieval using associative manufacturing features. J. Intell. Manuf. **14**, 183–195 (2003)
10. J. Jiao, L. Zhang, Y. Zhang, S. Pokharel, Association rule mining for product and process variety mapping. Int. J. Comput. Integr. Manuf. **21**(1), 111–124 (2008)
11. T.A.A. Kumar, A.S. Thanamani, Techniques to enhance productivity in manufacturing, Int. J. Eng. Sci. Invent. **2**(7), 9–13 (2013)
12. A.Y. Dorogov, A.S. Kharkovsky, Implementation data mining production model in analytic database, in *IEEE*, St. Petersburg, Russia, 2016
13. A. Bettacchi, A. Polzonetti, B. Re, Understanding production chain business process using process mining: A case study in the manufacturing scenario, in *Processing, Lecture Notes in Business Information*, vol. 249 (Springer, Cham, 2016)

A Predictive Model for Heart Disease Diagnosis Using Fuzzy Logic and Decision Tree

Asim Kumar Pathak and J. Arul Valan

Abstract Heart disease is one of the preeminent causes of death among people all over the world. Heart disease diagnosis procedure is a major and critical issue for the medical practitioner. Diagnosing the heart disease with the predictive model will have a tremendous effect on health care. Researchers have developed many predictive models and expert system using machine learning and data mining techniques for the diagnosis of heart disease. In this study, we propose a predictive model for heart disease diagnosis using a fuzzy rule-based approach with decision tree. In the proposed work, we have obtained the accuracy of 88% which is statistically significant for diagnosing the heart disease patient and also outperforms some of the existing methods. The proposed model uses only eight attributes of a patient to diagnose the heart disease, which is also a major advantage of our proposed model.

Keywords Heart disease · Decision tree · Fuzzy inference system · Predictive model

1 Introduction

Heart disease has become one of the leading causes of death for people in most countries around the world. In India also, cardiovascular diseases have increased and become a major cause of mortality. 52% patient having cardiovascular diseases die before the age of 70 years in India [1]. The occurrence of heart disease depends upon or can be affected by many features. So, it might be difficult for doctors to quickly and accurately diagnose heart disease. Detrimental consequences might be

A. K. Pathak (✉)
Department of CSE, BIT Sindri, Jharkhand, India
e-mail: asimkrpathak@gmail.com

J. Arul Valan
Department of CSE, NIT Nagaland, Nagaland, India
e-mail: valanmspt@yahoo.co.in

© Springer Nature Singapore Pte Ltd. 2020
A. Elçi et al. (eds.), *Smart Computing Paradigms: New Progresses and Challenges*,
Advances in Intelligent Systems and Computing 767,
https://doi.org/10.1007/978-981-13-9680-9_10

caused by poor clinical decisions. Therefore, the necessity of employing computerized technologies occurs to aid doctors to diagnose heart disease faster with higher accuracy.

In the last few decades, researchers have developed various machine learning and data mining techniques for intelligent heart disease prediction [2]. Some of the classification techniques used for heart disease prediction are neural networks, linear regression, support vector machine, Naive Bayes, decision tree, etc. Recently, many soft computing techniques have also been proposed for heart disease diagnosis.

In this study, we have developed a predictive model for diagnosing heart disease by combining decision tree and fuzzy rule-based approach and compared it with some of the existing methods. The results showed that the developed system outperforms some of the existing models in terms of accuracy.

2   Related Work

Various solutions have been suggested for heart disease prediction using different techniques. Anooj [3] proposed a clinical decision support system for the diagnosis of heart disease, where weighted fuzzy rules were used to build the system using Mamdani fuzzy inference system. This method achieved an accuracy of 62.3%. Bashir et al. [4] presented an ensemble classifier method for heart disease prediction where five heterogeneous classifiers such as Naive Bayes, decision tree based on Gini index, decision tree based on information gain, memory-based learner, and support vector machine were used. This method has achieved an accuracy of 88.52%. Long et al. [5] proposed a heart disease diagnosis system using rough set-based attribute reduction and interval type-2 fuzzy logic system (IT2FLS). The proposed approach faces several limitations like if a number of attributes are huge, attribute reduction is unmanageable, and again the training of interval type-2 fuzzy logic system by chaos firefly and genetic hybrid algorithms is quite slow. Pal et al. [6] proposed a fuzzy expert system for coronary artery disease (CAD) screening, where rules were formulated from the doctors, and fuzzy expert system approach was taken to deal with uncertainty present in the medical domain. Khatibi and Montazer [7] proposed a fuzzy-evidential hybrid inference engine for coronary heart disease risk assessment. They achieved 91.58% accuracy rate for its correct prediction. Weng et al. [8] used different types of neural network classifiers, and they built an ensemble classifier, which includes individual classifiers and solo classifier. They investigated the performance of those classifiers. They found the average accuracy of 82 and 84% for solo classifier and ensemble classifier, respectively, when five individual classifiers were used.

3 Material and Method

3.1 Dataset

The dataset [9] is collected from UCI machine learning repository. This dataset provides the results of various medical tests performed on a patient and also the diagnosis of presence or absence of heart disease on the patients. This database contains 14 attributes and 303 examples of patient. Out of 14 attributes, 13 attributes are medical test results and 1 attribute is the result of the diagnosis of heart disease, i.e., absence or presence of heart disease. The description of the dataset is given in Table 1.

3.2 Proposed Algorithm

The detailed explanation of the proposed algorithm is given in the following subsection. Figure 1 shows the flowchart of the proposed algorithm.

Table 1 Dataset description

Name	Type	Description
Age	Continuous	Age in years
Sex	Discrete	1 = male, 0 = female
Cp	Discrete	Chest pain type:
		1 = typical angina, 2 = atypical angina,
		3 = non-anginal, 4 = asymptomatic
Trestbps	Continuous	Resting blood pressure (in mm Hg)
Chol	Continuous	Serum cholesterol in mg/dl
Fbs	Discrete	Fasting blood sugar <120 mg/dl:
		1 = true, 0 = false
Restecg	Discrete	Resting electrocardiographic results:
		0 = normal, 1 = having ST-T wave abnormality,
		2 = left ventricular hypertrophy
Thalach	Continuous	Maximum heart rate achieved
Exang	Discrete	Exercise induced angina: 1 = yes, 0 = no
Old peak	Continuous	ST depression induced by exercise relative to rest
Slope	Discrete	The slope of the peak exercise segment:
		1 = up sloping, 2 = flat, 3= down sloping
Ca	Discrete	Number of major vessels (0–3) colored by fluoroscopy
Thal	Discrete	3 = normal, 6 = fixed defect, 7 = reversible defect
Diagnosis	Discrete	Classes: 0 = healthy, 1 = disease

Fig. 1 Flowchart of the
proposed algorithm

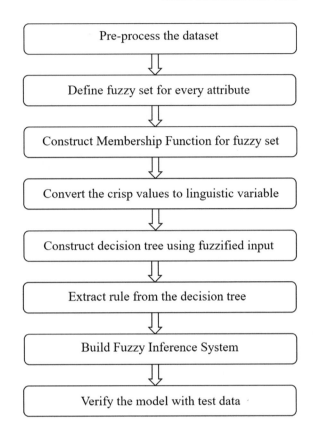

Fig. 1 Flowchart of the proposed algorithm

3.2.1 Preprocessing of the Dataset

Before using the dataset, it is preprocessed to remove the rows which have missing values. In our work, three rows which had missing values were removed. Then, the dataset is divided into two parts, training data and test data. Training data is used to build the model, and then, test data is used to verify the correctness of the constructed model.

3.2.2 Define Fuzzy Set for Every Attribute

In this step, we defined fuzzy sets for the attributes. Fuzzy sets are sets whose elements have degrees of membership [10]. Fuzzy sets defined for Age, Cholesterol, and Blood Pressure are given here: Age: young, middle-aged, old, and very old; Cholesterol: low, medium, and high; Blood pressure: low, medium, high, and very high.

3.2.3 Construct Membership Function for the Fuzzy Sets

For every defined fuzzy set, membership functions (MFs) have to be defined. There are a number of MFs available such as triangular, trapezoidal, Gaussian, bell-shaped, and sigmoidal. [10]. We have used triangular and trapezoidal MFs, which are best suited for our work. With the help of experts' knowledge, the ranges of different MFs are defined. MFs of Age and Blood Pressure are shown in Figs. 2 and 3, respectively.

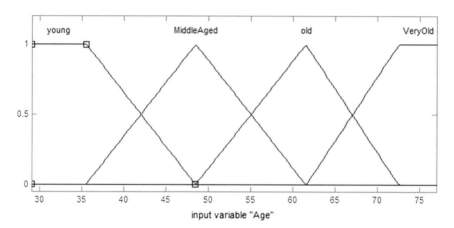

Fig. 2 MF for Age

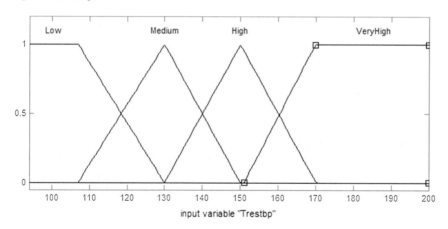

Fig. 3 MF for Blood Pressure

3.2.4 Convert Crisp Values to Linguistic Variable

In this step, using the MF, degree of membership of each value in the training dataset is obtained. Based on their degree of membership, the crisp values are converted to linguistic variables. Figure 4 shows a part of the original dataset and its conversion to linguistic variables.

3.2.5 Construct Decision Tree Using the Fuzzified Input

In this step, using the training dataset obtained in the previous step which contains the linguistics values, the output class decision tree is built, which consists of a root node representing the top node of the tree, leaf node representing the class label, internal nodes representing the attribute, and branches representing each possible value of the attribute node from which they originate [11]. For this purpose, we have used WEKA tool in our work.

3.2.6 Extract Rules from the Decision Tree

In this step, rules are extracted from the decision tree by traversing the tree from the root node to every leaf node [11]. Same number of rules equal to the number of leaf nodes can be extracted from each tree.

Original data

1	age	trestbp	chol	sex	cp	fbs	restecg	thalach	exang	oldpeak	slope	ca	thal	disease
2	63	145	233	1	1	1	2	150	0	2.3	3	0	6	0
3	67	160	286	1	4	0	2	108	1	1.5	2	3	3	2
4	67	120	229	1	4	0	2	129	1	2.6	2	2	7	1
5	37	130	250	1	3	0	0	187	0	3.5	3	0	3	0
6	41	130	204	0	2	0	2	172	0	1.4	1	0	3	0
7	56	120	236	1	2	0	0	178	0	0.8	1	0	3	0
8	62	140	268	0	4	0	2	160	0	3.6	3	2	3	3
9	57	120	354	0	4	0	0	163	1	0.6	1	0	3	0
10	63	130	254	1	4	0	2	147	0	1.4	2	1	7	2

Data converted to lunguistic variable

1	age	trestbp	chol	sex	cp	fbs	restecg	thalach	exang	oldpeak	slope	ca	thal	disease
2	old	high	med	m	typ	>120	hyper	med	no	med	down	c0	fix	no
3	v-old	vhigh	high	m	asym	<120	hyper	low	yes	med	flat	c3	norm	yes
4	v-old	med	med	m	asym	<120	hyper	med	yes	med	flat	c2	rev	yes
5	young	med	high	m	n-ang	<120	norm	high	no	high	down	c0	norm	no
6	young	med	med	f	atyp	<120	hyper	high	no	med	up	c0	norm	no
7	old	med	med	m	atyp	<120	norm	high	no	low	up	c0	norm	no
8	old	high	high	f	asym	<120	hyper	high	no	high	down	c2	norm	yes
9	old	med	high	f	asym	<120	norm	high	yes	low	up	c0	norm	no
10	old	med	high	m	asym	<120	hyper	med	no	med	flat	c1	rev	yes

Fig. 4 A part of original data and its conversion to linguistic variables

3.2.7 Construct the Fuzzy Inference System

The rules and MFs designed in the previous steps are now fed to a fuzzy inference system (FIS). We are using Mamdani inference system [12] in our work, which is the most commonly seen fuzzy methodology. We have MATLAB fuzzy logic toolbox to implement the FIS in our work.

3.2.8 Verify the Model with Test Data

In this step, test data is used to verify the correctness of the constructed model. The input attributes of test data are given as the input to the FIS, and then, the output from the FIS is compared with the output we already know for the test data.

4 Results and Discussion

We have used MATLAB and WEKA for implementing our proposed model. MATLAB fuzzy logic toolbox is used for building the fuzzy inference system, and WEKA j48 classifier is used for building the C 4.5 decision tree.

There are totally 300 patient data in the dataset which is divided into two sets, training set and test set. Training set consists of 80% of the total data, i.e., 240 number of data, and test set contains 20% of total data, i.e., 60 number of data. From the training data, three different decision trees with different pruning level are constructed and the rules from each tree are extracted. In WEKA J48 classifier, pruning of a decision tree depends on the confidence factor. We have used three different confidence factors 0.5, 0.4, and 0.2 for three different trees. The extracted rules are then fed to the FIS and tested with test data. For all three experiments, their accuracy, sensitivity, and specificity were calculated. The first tree with confidence factor 0.5 produced 42 rules, 13 attributes, 73% accuracy, 54% sensitivity, and 93% specificity. The second tree with confidence factor 0.4 produced 28 rules, 8 attributes, 75% accuracy, 58% sensitivity, and 93% specificity. The third tree with confidence factor 0.2 produced 21 rules, 7 attributes, 78% accuracy, 64% sensitivity, and 93% specificity. From these experiments, we have found that decision tree with confidence factor 0.2 has given the best result among all. So, for all the next experiments, we have created the decision trees using confidence factor 0.2 only.

To estimate better accuracy k-fold cross-validation is performed. According to this technique, the original data is divided into k subsets of data. Then, for every validation, a single subset is used as the test data and the remaining subsets all together form the training data. This procedure is repeated until all the subsets of data are utilized as test data. We have chosen $k = 5$ to analyze the proposed system. The results are shown in Table 2 and Fig. 5.

Table 2 Results of fivefold validation

No. of fold	No. of attributes	No. of rules	Accuracy	Sensitivity	Specificity
1	8	28	0.78	0.72	0.82
2	8	32	0.88	0.80	0.97
3	7	26	0.75	0.75	0.95
4	9	23	0.82	0.68	0.96
5	7	21	0.78	0.58	0.93

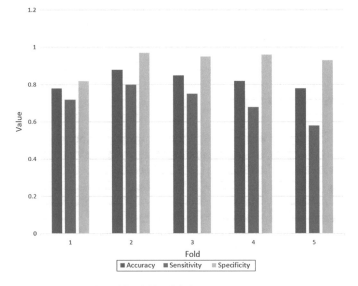

Fig. 5 Graphical representation of fivefold validation

From the fivefold validation, it is observed that the second fold gives us the better results, which used 8 attributes and generated 32 rules, and the highest accuracy of 88% is achieved.

5   Comparative Analysis

A comparative analysis is performed to find the efficiency of the proposed system. For comparison, we make use of the system proposed by Anooj [3], who developed a clinical decision support system of heart disease using weighted fuzzy rules, Bashir et al. [4], who developed an ensemble classifier for heart disease prediction, and Weng et al. [8], who used neural network classifiers for heart disease prediction (Table 3).

Table 3 Comparison with existing model

Work	Accuracy	Sensitivity	Specificity
Anooj [3]	0.62	0.76	0.44
Bashir et al. [4]	0.86	0.85	0.90
Weng et al. [8]	0.83	–	–
Proposed system	0.88	0.80	0.97

6 Conclusion

We have proposed a predictive model for heart disease diagnosis which can aid the doctors to diagnose heart disease easily and accurately. In our proposed method, the following results have been achieved. The best accuracy of 88% is obtained, which is statistically significant for diagnosing the heart disease patient and also outperforms some of the state-of-the-art methods. The best performance results have achieved with sensitivity 80%, specificity 97%, and accuracy 88%. Most of the previous researches used 13 attributes for prediction of heart disease, but this proposed model reduces the number of attributes from 13 to 8. Hence, a less number of tests are needed to perform on the patient, which is a major advantage of the proposed method. Further, this research may be extended to improve the accuracy, sensitivity, and specificity.

References

1. D. Prabhakaran, P. Jeemon, A. Roy, Cardiovascular diseases in India: current epidemiology and future directions. Circulation **133**(16), 1605–1620 (2016)
2. V. Krishnaiah, G. Narsimha, N.S. Chandra, Heart disease prediction system using data mining techniques and intelligent fuzzy approach: a review. Int. J. Comput. Appl. **136**(2), 43–51 (2016)
3. P.K. Anooj, Clinical decision support system: risk level prediction of heart disease using weighted fuzzy rules. J. King Saud Univ. Comput. Inf. Sci. **24**(1), 27–40 (2012)
4. S. Bashir, U. Qamar, F.H. Khan, M.Y. Javed, MV5: a clinical decision support framework for heart disease prediction using majority vote based classifier ensemble. Arab. J. Sci. Eng. **39**(11), 7771–7783 (2014)
5. N.C. Long, P. Meesad, H. Unger, A highly accurate firefly based algorithm for heart disease prediction. Expert Syst. Appl. **42**(21), 8221–8231 (2015)
6. D. Pal, K.M. Mandana, S. Pal, D. Sarkar, C. Chakraborty, Fuzzy expert system approach for coronary artery disease screening using clinical parameters. Knowl. Based Syst. **36**, 162–174 (2012)
7. V. Khatibi, G.A. Montazer, A fuzzy-evidential hybrid inference engine for coronary heart disease risk assessment. Expert Syst. Appl. **37**(12), 8536–8542 (2010)
8. C.H. Weng, T.C.K. Huang, R.P. Han, Disease prediction with different types of neural network classifiers. Telematics Inform. **33**(2), 277–292 (2016)
9. UCI machine learning repository, http://www.ncbi.nlm.nih.gov
10. G. Klir, B. Yuan, *Fuzzy Sets and Fuzzy Logic*, vol. 4 (Prentice Hall, New Jersey, 1995)

11. R. Lior, *Data Mining with Decision Trees: Theory and Applications*, vol. 81 (World Scientific, Singapore, 2014)
12. E.H. Mamdani, S. Assilian, An experiment in linguistic synthesis with a fuzzy logic controller. Int. J. Man Mach. Stud. **7**(1), 1–13 (1975)

Self-calibrating Thermocouple Using Neural Network Algorithm for Improved Sensitivity and Fault Identification

K. V. Santhosh

Abstract Automated calibration of temperature measurement technique is proposed in this work. The objective of the work is to design a technique which will be able to automatically calibrate the temperature sensor (thermocouple) output to obtain higher sensitivity and can also detect fault in sensor if any. The signal from the thermocouple is amplified using an instrumentation amplifier. The output of instrumentation amplifier is acquired on to the system using a general-purpose voltage data acquisition card (USB 6008). Based on user-defined range, the sensor output is calibrated to produce an output which has the highest sensitivity using neural network algorithms. Designed support vector algorithm is also trained to identify fault in sensor. The trained system is tested for evaluating its performance with both simulated and practical data. Results produced show successful achievement of set objective.

Keywords Calibration · Intelligence · Support vector machine · Temperature · Thermocouple

1 Introduction

Temperature monitoring system is available in almost all kind of processes loop. Like in the case of dairy industries, bakers, pharmaceuticals, cement manufacturing, glass manufacturing, dye making, etc., the temperature needs to be maintained at a particular set point to derive the product with desired quality. Inaccurate measurement of temperature would lead to malfunctioning of process, thereby hampering product quality. Temperature measurement system commonly would consist of temperature sensor followed by signal conversion and conditioning circuit. Purpose of signal conversion and conditioning circuits would be to convert the signal from one form to another along with performing functions like zero adjustment and gain adjust-

K. V. Santhosh (✉)
Department of Instrumentation and Control Engineering, Manipal Institute of Technology, Manipal Academy of Higher Education, Manipal, India
e-mail: kv.santhu@gmail.com

© Springer Nature Singapore Pte Ltd. 2020
A. Elçi et al. (eds.), *Smart Computing Paradigms: New Progresses and Challenges*,
Advances in Intelligent Systems and Computing 767,
https://doi.org/10.1007/978-981-13-9680-9_11

ment. Signal conditioning circuit may involve the functionality of calibration, or an additional circuit may sometime be used for calibration. Various kinds of temperature sensors are available. Sensors are considered based on a set of conditions like range of working, place of measurement, sensitivity, accuracy, response time, physical dimension, and economic issues. A detailed discussion is included in the next paragraph indicating the available types of sensor, characteristics analysis, and limitations.

A temperature sensor based on the splicing of a core offset multi-mode fiber with two single-mode fibers is proposed and demonstrated experimentally in [1]. Measurement of transient high temperature has been done using a thin-film thermocouple (TFTC) on printed circuit board (PCB) in [2]. Metrological features that have to be evaluated of various Raman distributed temperature sensing techniques (Raman-DTS) have been defined in [3]. An optical fiber-folded distributed temperature sensor based on Raman backscattering has been discussed in [4]. A temperature sensor based on an innovative hybrid resistive thread has been deliberated in [5]. In [6], a temperature sensor with peanut flat-end reflection structure has been discussed. To achieve thermal protection in Zynq system on chip technology, a fast relative digital temperature sensor has been implemented in [7]. These reported works are discussed on different temperature sensing technique available for measurement. From the reported works, it is seen that thermocouple-based temperature measurement is one of the widely used techniques because of its characteristics.

On further analysis of literature, it was seen that many researchers have worked on performance improvement of the temperature measurement technique. Based on the theory of heat conduction, a temperature sensor has been used in [8] to monitor bridge scour. A high-sensitivity, wide dynamic range differential temperature sensor has been discussed in [9]. A Sagnac-interferometer-based high-sensitivity temperature sensor with photonic crystal fibers has been discussed in [10]. A microfiber knot resonator-based high-sensitivity temperature sensor has been proposed in [11]. In [12], the effects of sensor coatings on the measurement of body temperature have been quantified. A smart temperature sensor based on resistive sensing and dual-slope ADC has been discussed in [13]. Fabrication and performances of temperature and pH sensors made of graphenic materials have been described in [14]. Responses of pressure and temperature were recorded with a change in luminescence intensities in [15]. In [16], the extension of temperature sensor linearity is carried out by using support vector machine is reported. From the detailed analysis of reported work, it can be understood that many researchers have worked on performance improvement in terms of increased linearity and sensitivity. But these methods involved changes incorporated at the sensor stage by modifying the design or materials. But these are economically not feasible and will not be of use to existing setup. Other modifications suggested are by repeated calibration, which is time consuming and involves manpower.

Few reported works have also been discussed on interdependence of temperature parameters and its need to measure and isolate the measurement. A double differential temperature compensation method has been discussed in [17] using a non-contact Fiber Bragg grating (FBG) vibration sensor. A fuzzy virtual temperature sensor has been proposed for an irradiative enclosure in [18]. In [19], for monitoring nitrate in real-time application, a graphene sensor has been employed and its advantages have been discussed. The distribution of electric potential and detection of temperature has been discussed in [20] using a combined temperature-potential probe. A flexible temperature with humidity insensitivity based on ion channel has been discussed in [21]. Simultaneous strain and temperature sensing have been performed in [22] based on polarization maintaining fiber and multi-mode fiber. A low-power wireless sensor network has been designed using two thermocouples in [23]. From the above study, it makes it quite important to study the temperature sensors. Implementation of an auto-calibration technique would be needed along with the identification of faults in sensors. In view of all the above reasons, an attempt is made to understand the characteristics of temperature sensor (thermocouple) over its full range and design suitable calibration technique for the ranges as specified by the user. Further, the designed process should be able to identify the faults in it if any.

2 Experimental Description

A thermocouple is an electrical device consisting of two dissimilar electrical conductors forming junctions to measure changes in temperature. In the proposed work, we use type K thermocouple. The output produced by thermocouple is in the ranges of few millivolts; an instrumentation amplifier is used to amplify the signal strength. The signal from the instrumentation amplifier is acquired on to personal computer for further processing using a data acquisition card (USB 6008).

3 Problem Statement

The temperature sensor once connected with the signal conversion and conditioning circuit is tested by subjecting it to changes in temperature. From the input–output characteristics, it can be seen that the full-scale sensitivity obtained is 2.93 mV/°C for the temperature range from −200 to 1200 °C. If similar system is used to measure temperatures with lower ranges like 0–200 °C, the output produced will be of the range from −0.16 to 0.22 V. Considering the fact that only partial range of input is used, the output could have always been taken full scale, thereby improving the sensitivity. For achieving this, the process of calibration needs to be repeated again, which is time consuming and economically not profitable.

4 Problem Solution

In this stage, an algorithm is designed for attaining the set objective. The first process is to acquire the input from the data acquisition card. LabVIEW software is used for the purpose of data processing and monitoring. The front panel of the designed tool is as shown in Fig. 1, and the front panel consists of two user-defined controls to feed the minimum and maximum measurable ranges. Two indicators are placed to show the calibrated voltage corresponding to measured temperature and the other one to indicate measured temperature. One indicator is placed to show the status of sensor where color indicator green represents functional sensors and red indicting faults sensor. Support vector algorithms are used for this purpose [24]. In the current work, regression problem is solved, and the objective function is to train the output of thermocouple signal conversion circuit output such that it auto-calibrates to produce desired output having maximum linearity and sensitivity. SVM regression analysis can be represented by

$$h(y) = f(x) + E \tag{8}$$

where

$h(y)$ Target vector

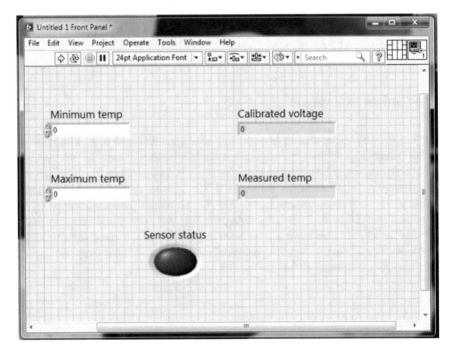

Fig. 1 Front panel of proposed technique

$f(x)$ Predictor
ε Error/noise

The next function is to identify the sensor faults like open-circuit fault, short-circuit fault, and drift fault. In open-circuit fault, the condition would be to check whether the output is normally zero. Output is checked for multiple cycles if value appears to be zero regressively. Secondly, for short-circuit fault, the current value is checked such nearly three cycles the value of output is checked for saturation voltage. For checking the drift error, impulse signal is subjected at specific intervals and output derived for those values are compared to identify these faults.

5 Results and Analysis

The designed system once trained is tested with practical data under different ranges of measurement. For testing, the measurement was carried out at varying ranges of measurement like in case 1 (24–50 °C), case 2 (50–70 °C), case 3 (32–100 °C), case 4 (28–80 °C), case 5 (40–90 °C), case 6 (24–90 °C), etc. These specimens were expected to get calibrated to the value of 1–5 V for the ranges specified. But in case of the earlier system, the calibration was carried over for the range from −200 to 1200 °C. Results obtained from each of the auto-calibrated system are indicated in Fig. 2.

From the plots in Fig. 2, it is clear that the proposed measurement system has been able to get calibrated over the ranges specified by the user automatically. The

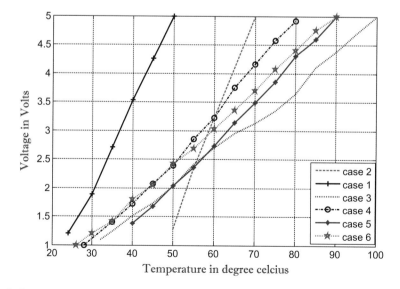

Fig. 2 Input–output characteristics of temperature measurement

results show that the proposed technique is able to tune the gain and offset such that it always maintains maximum sensitivity.

6 Conclusion and Discussion

Temperature measurement is a very vital industrial process in most of the industries. Maintaining the best performance from these measurement systems will always be the aim of process engineers. The best-performing measurement system is often expected to have high accuracy, precision, sensitivity, and bandwidth. The proposed system was designed with an intention to achieve higher sensitivity in measurement automatically. Test results show the successful implementation of desired objectives.

References

1. X. Fu et al., A temperature sensor based on the splicing of a core offset multi-mode fiber with two single mode fiber. Optoelectron. Lett. **11**(6), 434–437 (2015)
2. W.L. Wang et al., PCB-integrated thin film thermocouples for transient temperature measurement. Electron. Lett. **52**(13), 1140–1141 (2016)
3. G. Failleau et al., A metrological comparison of Raman-distributed temperature sensors. Measurement **116**, 18–24 (2018)
4. Z. Wang et al., An optical fiber-folded distributed temperature sensor based on Raman backscattering. Opt. Laser Technol. **93**, 224–227 (2017)
5. R. Polanský et al., A novel large-area embroidered temperature sensor based on an innovative hybrid resistive thread. Sens. Actuators A **265**, 111–119 (2017)
6. Y. Li et al., High-temperature sensor based on peanut flat-end reflection structure. Opt.-Int. J. Light Electron Opt. **148**, 293–299 (2017)
7. C.-A. Lefebvre, J.L. Montero, L. Rubio, Implementation of a fast relative digital temperature sensor to achieve thermal protection in Zynq SoC technology. Microelectron. Reliab. **79**, 433–439 (2017)
8. Y. Ding et al., A new type of temperature-based sensor for monitoring of bridge scour. Measurement **78**, 245–252 (2016)
9. E. Vidal et al., Differential temperature sensor with high sensitivity, wide dynamic range and digital offset calibration. Sens. Actuators A **263**, 373–379 (2017)
10. Q. Liu, S.-G. Li, H. Chen, Enhanced sensitivity of temperature sensor by a PCF with a defect core based on Sagnac interferometer. Sens. Actuators B **254**, 636–641 (2018)
11. J. Li et al., A high sensitivity temperature sensor based on packaged microfibre knot resonator. Sens. Actuators A **263**, 369–372 (2017)
12. S. Snyder, P.J.S. Franks, Quantifying the effects of sensor coatings on body temperature measurements. Anim. Biotelemet. **4**(1), 8 (2016)
13. L. Zou, A resistive sensing and dual-slope ADC based smart temperature sensor. Analog Integr. Circ. Sig. Process **87**(1), 57–63 (2016)
14. P. Salvo et al., Temperature and pH sensors based on graphenic materials. Biosens. Bioelectron. **91**, 870–877 (2017)
15. S. Sano et al., Temperature compensation of pressure-sensitive luminescent polymer sensors. Sens. Actuators B **255**, 1960–1966 (2018)

16. K.V. Santhosh, B.K. Roy, Soft calibration technique with SVM for intelligent temperature measurement using thermocouple, in *Proceedings International Conference on Advances in Electronics, Computers, and Communications*, India, 2014

17. T. Li et al., A non-contact FBG vibration sensor with double differential temperature compensation. Opt. Rev. **23**(1), 26–32 (2016)

18. D. Mehrabi et al., A fuzzy virtual temperature sensor for an irradiative enclosure. J. Mech. Sci. Technol. **31**(10), 4989–4994 (2017)

19. M.E.E. Alahi et al., A temperature-compensated graphene sensor for nitrate monitoring in real-time application. Sens. Actuators A **269**, 79–90 (2018)

20. L. Bühler et al., Development of combined temperature–electric potential sensors. Fusion Eng. Des. (2017)

21. J.-S. Kim, K.-Y. Chun, C.-S. Han, Ion channel-based flexible temperature sensor with humidity insensitivity. Sens. Actuators A (2018)

22. R. Xing et al., Simultaneous strain and temperature sensor based on polarization maintaining fiber and multimode fiber. Opt. Laser Technol. **102**, 17–21 (2018)

23. V. Markevicius et al., Two thermocouples low power wireless sensors network. AEU-Int. J. Electron. Commun. **84**, 242–250 (2018)

24. N. Cristianini, J. Shawe, *An Introduction to Support Vector Machines and Other Kernel-Based Learning Methods* (Cambridge University Press, 2000)

SGMM-Based Modeling Classifier for Punjabi Automatic Speech Recognition System

Virender Kadyan and Mandeep Kaur

Abstract A baseline ASR system does not perform better due to improper modeling of training data. Training of system through conventional HMM technique faced the issue of on or near manifold in data space. In this paper, Hybrid SGMM-HMM approach is compared with baseline GMM-HMM technique on Punjabi continuous simple sentences speech corpus. It examined the hybridized HMM technique: SGMM-HMM to overcome the problem of sharing of state parameter information throughout the training and testing of system. The system testing is performed on Kaldi 4.3.11 toolkit using MFCC and GFCC approaches at the front end of the system. The result obtained on SGMM-HMM modeling technique generates an improvement of 3–4% over GMM-HMM approach. The experiments are performed on real environment dataset.

Keywords MFCC · GFCC · SGMM-HMM · GMM-HMM

1 Introduction

Extraction and modeling of discriminative information have been the active area of research in ASR technology [1]. After retrieving of meaningful information through the first step of feature extraction to represent them into a compact form. The MFCC and GFCC are two dominant approaches in the recent trend in speech recognition systems that showcase better results [2]. The performance of speech recognition system is greatly influenced by auditory the environment that needs to be analyzed either in training or testing of the system. In the last four decades, various statistical approaches such as HMM are adopted that helps in resolving the issue of training over the large dataset. However, most of the researchers have been tried to hybridize

V. Kadyan · M. Kaur (✉)
Department of Computer Science and Engineering, Chitkara University Institute of Engineering and Technology, Chitkara University, Punjab, India
e-mail: kaurmandeep@chitkara.edu.in

V. Kadyan
e-mail: virender.kadyan@chitkara.edu.in

© Springer Nature Singapore Pte Ltd. 2020
A. Elçi et al. (eds.), *Smart Computing Paradigms: New Progresses and Challenges*,
Advances in Intelligent Systems and Computing 767,
https://doi.org/10.1007/978-981-13-9680-9_12

HMM with GMM and DNN modeling classifiers that perform up to great extent on the large or medium training corpus.

In this paper, we present an approach of SGMM-HMM for acoustic modeling on two front-end approaches such as GFCC and MFCC. The training and testing of the system are performed using an open-source toolkit that is Kaldi [3]. The system testing is performed individually on GFCC and MFCC with SGMM-HMM and GMM-HMM classifiers. The comparative study is shown by among two front-end and two back-end approaches. The rest part of the paper is organized as follows. The related work is described in Sect. 2. Section 3 explains an overview of hybrid modeling approaches. Section 4 briefly describes the SGMM-based modeling approach for Punjabi automatic speech recognizer. Finally, Sects. 5 and 6 include our results and conclusion.

2 Related Work

In recent papers, [4] has developed the Arabic automatic speech recognition using MFCC technique and tried to achieve a word accuracy of 98.01%. In [5], a Hindi ASR system is presented using HTK toolkit and conceived a word accuracy of 94.63% on 30 Hindi datasets collected from eight speakers. Small vocabulary Punjabi automatic speech recognition for continuous speech using triphone-based acoustic modeling is presented that produced an overall recognition of 82.18% in [6]. In [7], a Punjabi-isolated system for refining of acoustic information is performed using two-hybrid HMM classifiers: GMM-HMM and DE+HMM with the help of MFCC approaches that achieved a word accuracy of 84.2%. The work tried to extend on mismatch train and test condition using different combination of front-end approaches such as MFCC, PLP, RASTA, and MF-PLP on hybrid modeling classifier such as DE+HMM, GA+HMM, and HMM [8]. The speech recognition using ANN and swarm intelligent approaches is described in [9] on isolated word database in two different environments clean and noisy. The system obtained better results with ANN trained system using PPO, CSO, and backpropagation approaches. A GFCC-based robust speech recognition system is developed in [10] using an integrated feature that yields better results on conventional acoustic features. In [11], a discriminative approach of bmmie is tried to be examined on 270 h of Punjabi speech corpus and attain an accuracy of 79.2% (MFCC with 2 g), 79.2% (MFCC with 3 g), 79.2% (PLP with 2 g), and 79.2% (PLP with 3 g), on different n-gram modeling stages.

3 Approaches for Hybridization of HMM Classifier

The classification of acoustic information the extracted feature vectors are employed as an input. The classification process further helps the better recognition of the test sample of a spoken utterance. State of art speech system depends on HMM technique for acoustic modeling that employed probability distribution of an acoustic vector. The increment of acoustic information is tried to represent them in a better way over states of HMM. The HMM is hybridized with GMM or SGMM to boost the performance recognition. GMM is integrated on HMM where each state space of an input speech is modeled using multivariate GMM. These states store the parameter of GMM but entire parameters are not employed for overall model. The state contains the feature vector information that matches to the value the dimension of an extracted feature vector. The global mapping is shared between the state vector corresponding to its mean and weight of a GMM states. The learning of weight plays a key role in the model. These parameters are learned through Baum–Welch algorithm. The GMM system put the constrained on adoption of an allocated number of a mixture of its multivariate Gaussian. It helps in the modeling of each state level that does not allow later sharing of state parameter information.

$$P\left(\frac{x}{y}\right) = \sum_{i=1}^{I} W_{ji} N\left(x; \mu_{ji}; \sum_{i}\right) \qquad (1)$$

$$\mu_{ji} = M_i V_j \qquad (2)$$

$$W_{ji} = \frac{\exp W_i V_j}{\sum_{i=1}^{I} \exp W_i V_j} \qquad (3)$$

where j is the speech state, W_{ji} is the mixture weights, μ_{ji} mean and covariances $\sum i$ which are shared between states. M_i is globally shared parameter. This model is similar to tied–mixture because the Gaussian within each state difference in mean as well as weight. It also uses the full covariance than diagonal because full covariance does not sustainably change either the parameter count or the decoding speed.

4 System Overview

A basic architecture of SGMM-HMM-based Punjabi automatic speech recognition system is presented in Fig. 1. Initially, GFCC and MFCC feature vectors are extracted through static 13 and their delta or double delta feature vector in order to generate a total of 39 feature vectors. These feature vectors are used to train on context-dependent (triphone) and context-independent (monophone) state with the help of provided acoustic observations. Initially, monophone state is trained to train the triphone state that helps in training of GMM-HMM model. This state information is further employed by SGMM-HMM approach. The standard Kaldi open-source toolkit is adopted to train the SGMM-HMM and GMM-HMM models using the following steps:

1. Compute MFCC or GFCC features for input audio segments to generate the 13-dimensional feature vector.

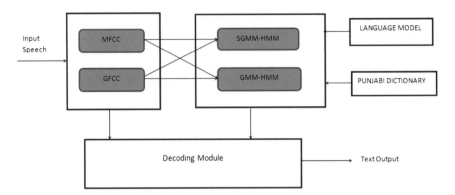

Fig. 1 Workflow of SGMM-HMM-based Punjabi ASR system

2. Compute CVNM starts for the frames.
3. Train the monophone model to produce graph.
4. Align the output of monophone model using delta and delta-delta's coefficients to produce its graph on triphone model.
5. Apply LDA and MLLT using triphone and delta-delta's alignments to produce data graph.
6. Compute train data using SAT technique on LDA+MLLT alignment.
7. If (GMM-mode == 1)
   Analyze the test data using step 4–step 6 combinations.

   else
   Analyze the test data using SGMM-HMM model.

5 Experiments

The proposed system is evaluated on medium vocabulary Punjabi continuous speech corpora. The clean data is initially used for training and testing of the proposed system with two hybrid classifiers SGMM-HMM and GMM-HMM. The system testing is performed with the help of two front-end approaches (MFCC and GFCC). The MFCC outperforms GFCC in clean environment corpus whereas GFCC performed better on SGMM-HMM approaches. The system is evaluated on two sets using the following ways:

- In the first set, parameters of SGMM (such as a number of Gaussian and number of leaves) are varied to tune them to a fixed value that generates better accuracy.
- In the second set, the comparative study of SGMM-HMM and GMM-HMM approaches is used. These classifiers are compared with two front-end approaches like GFCC and MFCC. The system testing is performed using two standard formulas of WER and WA:

$$\text{WER}(\%) = \frac{S + I + D}{N}$$
$$\text{WA}(\%) = 1 - \text{WER}$$

where S is substitution, I insertion, D deletion and N total number of spoken words.

5.1 Experiments with Gaussian Mixtures and Num Leaves in SGMM

Speaker adaption of hidden Markov model (HMMs) using the FMLLR method is implemented to improve the performance of the system on two different front-end techniques GFCC and MFCC. The analysis is obtained with acoustic modeling technique subspace Gaussian mixture model (SGMM) through a number of Gaussian that is varied from (7000–12,000). The system performed with a very low error rate with GFCC at 7000 gaussian value as shown in Fig. 2a.

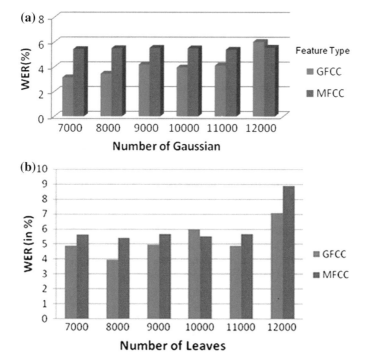

Fig. 2 a WER evaluation on varying number of leaves in SGMM. **b** WER evaluation on varying number of Gaussian in SGMM

Table 1 WER evaluation on different acoustic modeling classifier with two front-end approaches

Feature type	Modeling classifier	I	S	D	WA
MFCC	GMM-HMM	35	31	358	96.80
	SGMM-HMM	44	23	290	97.20
GFCC	GMM-HMM	6	854	2374	95.41
	SGMM-HMM	6	592	2312	98.21

Secondly, number of leaves are also tried to be varied from 7000 to 12,000, respectively. The system obtained minimum word error rate at num leaves the value of 8000. It can be observed that as the number of leaves increases system accuracy decreases drastically in MFCC but GFCC output is varied slowly as depicted in Fig. 2b.

5.2 Experiments with Heterogeneous Modeling Classifier Using Two Front-End Approaches

Experiments are performed with different modeling classifier approach on two feature extraction approaches such as GFCC and MFCC. The system obtained minimum word error rate of 3.01% with GFCC approach on SGMM-HMM technique as shown in Table 1. The system obtained better results either in matched or in mismatched train and test conditions. Due to this overhead of the system with increases in training corpus is a decrease that affects its word accuracy with SGMM-HMM approach.

6 Conclusions

A study of two hybrid modeling classifiers is presented on two front-end approaches. As none of the proposed feature approach work efficiently on baseline modeling approaches such as HMM and GMM. The combinations of techniques yield better results with an improvement of 3–4%. To achieve an optimum result, different parameter of SGMM-HMM approach is varied such as a number of Gaussian mixture and number of leaves. The system obtained better results on mismatched trained and test condition with GFCC approach using SGMM-HMM classifier on continuous Punjabi speech corpora. Hence, the work can be extended in further by increasing the size of the vocabulary and can be tried to be implemented in real environment ASR system.

References

1. M.J. Alam, P. Kenny, D. O'Shaughnessy, Low-variance multitaper mel-frequency cepstral coefficient features for speech and speaker recognition systems. Cognit. Comput. **5**(4), 533–544 (2013)
2. Z. Wu, Z. Cao, Improved MFCC-based feature for robust speaker identification. Tsinghua Sci. Technol. **10**(2), 158–161 (2005)
3. D. Povey, The Kaldi speech recognition toolkit, in *IEEE 2011 Workshop on Automatic Speech Recognition and Understanding*. No. EPFL-CONF-192584. IEEE Signal Processing Society, 2011
4. B.A.Q. Al-Qatab, R.N. Ainon, Arabic speech recognition using hidden Markov model toolkit (HTK), in *International Symposium in Information Technology (ITSim), 2010*, IEEE, vol. 2 (2010)
5. K. Kumar, R.K. Aggarwal, A. Jain, A Hindi speech recognition system for connected words using HTK. Int. J. Comput. Syst. Eng. **1**(1), 25–32 (2012)
6. W. Ghai, N. Singh, Continuous speech recognition for Punjabi language. Int. J. Comput. Appl. **72**(14) (2013)
7. V. Kadyan, A. Mantri, R.K. Aggarwal, A heterogeneous speech feature vectors generation approach with hybrid hmm classifiers. Int. J. Speech Technol. **20**(4), 761–769 (2017)
8. V. Kadyan, A. Mantri, R.K. Aggarwal, Refinement of HMM model parameters for Punjabi automatic speech recognition (PASR) system. IETE J. Res. 1–16 (2017)
9. T. Mittal, R.K. Sharma Speech recognition using ANN and predator-influenced civilized swarm optimization algorithm. Turk. J. Elec. Eng. Comput. Sci. **24**(6), 4790–4803 (2016)
10. Y. Shao, An auditory-based feature for robust speech recognition, in *IEEE International Conference on Acoustics, Speech and Signal Processing, 2009*, ICASSP, IEEE (2009)
11. J. Guglani, A.N. Mishra, Continuous Punjabi speech recognition model based on Kaldi ASR toolkit. Int. J. Speech Technol. 1–6 (2018)
12. D. Povey, Subspace gaussian mixture models for speech recognition, in *IEEE International Conference on Acoustics Speech and Signal Processing (ICASSP)*, IEEE, 2010

Augmenting a Description Logic with Probability for Motion Patterns Within QSTR

Upasana Talukdar, Rupam Barua and Shyamanta M. Hazarika

Abstract Motion patterns are spatiotemporal. Human everyday spatiotemporal reasoning is predominantly qualitative; hence, a qualitative abstraction of motion patterns holds promise. Reasoning over qualitative models of spatiotemporal change is referred to as qualitative spatiotemporal reasoning (QSTR). Recognition of patterns within a QSTR framework involves reasoning over a multitude of entities. These entities interacting in numerous ways often lead to hierarchies and ample uncertainties. Herein lies the motivation of using a formal basis for the representation of large and diverse knowledge. Description logics (DLs), a family of knowledge representation formalism providing object-oriented representation with formal semantics, are a natural choice. However, traditional DLs delineate a delicate approach to handle uncertain features and relationships. A paradigm that combines a DL formalization with probability theory is required. Bayesian Networks (BNs) are the most widely used models to represent knowledge about an uncertain domain. This paper combines a DL with BN and proposes a formal, explicit qualitative knowledge representation formalism for recognition of motion patterns within an uncertain domain.

Keywords Description logic · Qualitative spatiotemporal reasoning · Bayesian Network · Motion patterns

U. Talukdar (✉)
Biomimetic & Cognitive Robotics Lab, Department of Computer Science & Engineering,
Tezpur University, Tezpur, India
e-mail: upat123@tezu.ernet.in

R. Barua
Department of Computer Science & Engineering, Jorhat Engineering College, Jorhat, India

S. M. Hazarika
Mechatronics and Robotics Lab, Department of Mechanical Engineering,
Indian Institute of Technology, Guwahati, India
e-mail: s.m.hazarika@iitg.ernet.in

© Springer Nature Singapore Pte Ltd. 2020
A. Elçi et al. (eds.), *Smart Computing Paradigms: New Progresses and Challenges*,
Advances in Intelligent Systems and Computing 767,
https://doi.org/10.1007/978-981-13-9680-9_13

1 Introduction

Objects in motion change space over time. Such a change of space often exhibits perceptible regularity. This kind of regular motion occurring over time is defined as a motion pattern. Motion patterns are spatiotemporal [1]. One of the major challenges within *Artificial Intelligence* is to learn motion patterns from spatiotemporal data streams. The task is difficult because of the necessity to not only track and detect objects but also resolve their spatiotemporal interactions. Nevertheless, human performs recognition of motion patterns with utmost ease. Human everyday spatiotemporal reasoning is predominantly qualitative [1, 2]. Knowledge representation and reasoning (KR & R) for interaction with the physical world is through qualitative abstractions. Qualitative approach to KR & R for high-level abstraction of motion patterns holds promise.

Desiderata of a Formal Basis for Motion Patterns: Motion patterns can be best understood from an example encountered in everyday road traffic. Figure 1 depicts road traffic scenario at three distinct intervals of time. Initially, the car B *follows* the truck A. Sometimes both move together. Finally, the truck A *follows* the car B. Given our background knowledge, we can very well see an *overtake*. Here, both 'follows' and 'overtake' are motion patterns. The example brings to light the following observations with regard to motion patterns within spatiotemporal data streams.

Fig. 1 Motion pattern of *Overtake* as observed in everyday road traffic. Bottom row depicts the initial scenario where car B *follows* truck A. Middle row shows both the truck A and the car B together. Top row shows final scenario where truck A *follows* car B

a. Background knowledge has a substantial role in the recognition of motion patterns.
b. Qualitative abstractions form the basis of representing motion patterns for everyday reasoning.
c. Motion patterns inherently have a hierarchical structure, and a pattern can be decomposed into its sub-patterns.

Herein lies the motivation of using a formal basis for the representation of qualitative knowledge for motion patterns. Description logics (DLs), a family of knowledge representation formalism providing object-oriented representation with formal semantics, are a natural choice. We use $\mathcal{STDL}$, a description logic-based QSTR framework [3]. However, traditional DLs delineate a delicate approach to handle uncertain features and relationships. A paradigm that combines a DL formalization with probability theory is required. Bayesian Networks (BNs) are the most widely used models to represent knowledge about an uncertain domain. This paper combines $\mathcal{STDL}$ with BN and proposes a formal, explicit Bayesian spatiotemporal DL ($\mathcal{BSTDL}$) for recognition of motion patterns within an uncertain domain.

2 Basic Knowledge Representation Formalisms

In this section, we briefly review the qualitative spatiotemporal representation and the spatiotemporal DL employed here. We describe the qualitative abstractions, which form the elementary representation formalism for the representation of motion patterns. $\mathcal{STDL}$ —the DL formalism that describes different ways of combining these basic spatiotemporal abstractions to form more complex representation structures is reviewed.

2.1 Qualitative Spatiotemporal Abstraction

For qualitative description of any motion pattern such as the one described in the above traffic scenario, distance, direction, and orientation are some of the crucial parameters.

Qualitative Directions: We use the qualitative direction algebra QD_8 for representation of the direction [4]. The direction relations (C_D) are defined w.r.t. an egocentric frame of reference (FOR). There are 12 base relations. These are: Same+, Same, Same-, LR+, LR, LR-, Opposite+, Opposite, Opposite-, RL+, RL, and RL-. Each element of (C_D) is a direction [4].

Qualitative Orientation: Qualitative orientation relations (C_O) are also defined w.r.t an intrinsic frame of reference. Twenty-eight base relations are defined. The base relations are: Front, Left, Right, Back, Front & Left, Front & Right, Back & Left, Back & Right, ExtendedFront, ExtendedLeft, ExtendedRight, ExtendedBack,

FrontOverlap, LeftOverlap, RightOverlap, BackOverlap, Front & LeftOverlap, Front & RightOverlap, Back & Left-Overlap, Back & RightOverlap, ExtendedOverlapIn-Front, ExtendedOverlapInLeft, ExtendedOverlapInRight, ExtendedOverlapInBack, AllRegion, AllRegionOverlap, Inside, and Equal. Each base relation is an orientation abstraction. For more details on the qualitative abstraction, see [3, 5].

Qualitative Spatiotemporal Change: Spatiotemporal relations combine the space and time abstraction for representation of motion event. Scenarios that change over time are assumed to comprise of *episodes*. Each *episode* defines a particular qualitative spatial relation of a primary object w.r.t. a reference object [3, 5].

2.2 $\mathcal{STDL}$—Spatiotemporal DL Framework

Basic ingredients of $\mathcal{STDL}$ are : 1. Logical symbols that includes *punctuation* ([,]), *connectives* ($\doteq$), and *concept forming operators* (ALL, AND, FILLS, MEETS, AT); 2. Non-logical symbols that consists of *concepts* (Spatial_Concept, $\mathcal{SC}$, Temporal_Concept,$\mathcal{TC}$, and Spatio_Temporal_Concept,$\mathcal{STC}$), *roles* (Primary_Object, R_D for Spatial Concept D_i and R_O for Spatial Concept O_i), and *constants* [3].

$\mathcal{STDL}$ knowledge base consists of an A-Box and a T-Box. A-Box includes role assertions (*RA*) and concept assertions. A-Box provides the atomic facts which are then used to encode T-Box. T-Box aims to reason and represent the semantics of motion sequences between the primary and reference object. For that two types of motion patterns are defined *fundamental pattern* (F= [AND [FILLS Primary_Object c][ALL R_i C_i]]) and *basic pattern* (B= [AND F_D F_O]; F_D is for the F for direction and F_O is the F for orientation) of the primary object w.r.t. reference object. The *basic patterns* of different episodes of a particular scenario constitute the T-Box [3]. The A-Box and T-Box together encode the $\mathcal{STDL}$ knowledge base [3].

3 DL + BN: A KR & R Formalism for Uncertain Domains

3.1 $\mathcal{L}$: Logical Bayesian Network

The proposed approach is established through the inclusion of the Bayesian Network within $\mathcal{STDL}$ to present probabilistic reasoning involving $\mathcal{STDL}$ concepts. The syntactic structure of such a network that combines a directed acyclic graph with roles and concepts of $\mathcal{STDL}$ is defined in Definitions 1 and 2.

Definition 1 Logical Network LN= $(V,\mathcal{E})$ is a directed acyclic graph where

1. V is the set of nodes which are either concepts (either direction $D_i \in C_D$ or orientation $O_i \in C_O$) or roles (R_D or R_O) hold by the primary object in the scenario w.r.t. the reference object, and

2. $\mathfrak{E}$ is the set of edges characterizing the conditional dependencies of the nodes represented by directed arrows such that $\mathfrak{E} \subseteq V \times V$.

Here, role (R_D or R_O) is the root node, and concepts hold by the primary object are the child nodes. If R_D is the root node, then its child nodes will be $D_i \in C_D$ and if R_O is the root node, then its child nodes will be $O_i \in C_O$.

Logical Network, $LN^1 = (V1, \mathfrak{E}1)$, is for a particular scenario which includes only those orientations or directions that are held by the primary object w.r.t. the reference object that is there in the scenario such that $V1 \subseteq V$ and $E1 \subseteq E$.

The concepts that makes up LN^1 are of two types:

1. **True Concept**: Any node C_i of LN^1 such that $C_i \in C_{O\,1}$ or $C_i \in C_{D\,1}$ for a particular scenario is referred as true concept ($\mathfrak{T}$). For true concepts, concept probability is defined.

   Concept Probability: The probability of true node, $\mathfrak{T}$, is termed as concept probability and is given $\alpha = o/m$, where o is the no. of occurrences of $\mathfrak{T}$ during the scenario and m is the number of episodes of the scenario.

2. **Pseudo-concept**: Any node C_i of LN^1 such that $C_i \notin C_{O\,1}$ or $C_i \notin C_{D\,1}$ of a particular scenario but is needed to construct the LN^1 for that scenario is termed as pseudo-concept ($\mathfrak{F}$).

Definition 2 A Logical Bayesian Network, $\mathfrak{L}$, for a particular scenario is

1. A Logical Network $LN^1 = (V1, \mathfrak{E}1)$ where $V1$ is the set of nodes that includes true concepts, $\mathfrak{T}$ along with the pseudo-concepts, $\mathfrak{F}$.
2. A Logical Network LN^1 in which each node is associated with a conditional probability table.

Proposition 1 *Given a scenario with roles, concepts, and n episodes E_1, E_2, E_3, E_n; A Logical Bayesian Network ,$\mathfrak{L}$, can always be constructed at E_n.*

Proof By Definition 1, Roles form the root node and true concepts, $\mathfrak{T}$, along with pseudo-concepts, $\mathfrak{F}$, are child nodes. The qualitative framework for orientation (defined in Sect. 3.1.2.) is started with four qualitative orientation relations—Front, Left, Right, and Back, i.e., Front $\cup$ Left $\cup$ Right $\cup$ Back $\rightarrow$ Orientation Relations. Hence, Front, Left, Right, and Back are the child nodes of R_O and forms Level 1 Nodes. They are further refined into ten relations like Front & Right which is the combination of Front and Right and hence the child node of the same and thus form the Level 2 Nodes. Similarly the others. In case of direction relation, role R_D is the root node. Here too, similar concept is followed where Same, LR, RL, and Back are the child nodes of R_D forming the Level 1 Nodes and so are the others. So, given a scenario with roles, concepts, and episodes , with this concept of parent–child relationship and their dependencies, Logical Network LN can be constructed. The conditional probability is constructed using histories and assigned to each node which constructs Logical Bayesian Network $\mathfrak{L}$. ∎

3.2 $\mathcal{BGTDL}$: A Bayesian Spatiotemporal DL

Syntax: $\mathcal{BGTDL}$ combines $\mathcal{L}$ with $\mathcal{GTDL}$. Let us consider a $\mathcal{BGTDL}$ knowledge base which consists of three components:

1. *Probabilistic A-Box* ($\mathbb{P}_{AB}$)—comprises probabilistic axioms ($\mathbb{P}_A$) and role assertion set, S_{RA}. $\mathbb{P}_A$ is of the form : CA^c: η where CA is of the form $(D_i(c), O_i(c))$ where $D_i \in C_{D1}$ and $O_i \in C_{O1}$, and $\eta = \alpha 1 \times \alpha 2$. Here, CA is the concept assertion of object c, $\alpha 1$ is the concept probability of D_i, and $\alpha 2$ is the concept probability of O_i.
2. *Probabilistic T-Box* ($\mathbb{P}_{TB}$) - To define *probabilistic T-Box*($\mathbb{P}_{TB}$), three types of probabilistic patterns ($\mathbb{P}$) need to be defined. $\mathbb{P}$ is actually Spatio_Temporal_Concepts or Spatial_Concepts associated with a positive probability.
3. *Logical Bayesian Network*($\mathcal{L}$)—as defined above.

Definition 3 A probabilistic fundamental pattern ($\mathbb{P}_F$) of an episode E_i is a Spatial _ Concept of the form - F : ξ where F is the fundamental pattern either for direction (F_D) or orientation (F_O) and ξ is the conditional probability of node from $\mathcal{L}$ which is a true concept,$\mathfrak{T}$ of E_i.

Definition 4 A probabilistic basic pattern (P_B) is a Spatial_Concept of the form: B_i : ρ where ρ is the probability of the basic pattern B_i for episode E_i. Each $\mathbb{P}_B$ consists of one probabilistic fundamental pattern for direction, (F_D : ξ_D) and another for orientation (F_O : ξ_O), i.e.,

$$B_i : \rho \Rightarrow [AND\ F_D : \xi_D\ F_O : \xi_O] : \rho$$

Definition 5 A probabilistic neighboring pattern ($\mathbb{P}_{NP}$) is a Spatio_Temporal_Concept of the form: $NP_{i,i+1}$: σ where σ is the probability of a neighboring pattern for two neighbor episodes E_i and E_{i+1}. Each probabilistic neighboring pattern consists of two probabilistic basic patterns, i.e.,

$$NP_{i,i+1} : \sigma \Rightarrow [ANDMEETS E_i E_{i+1}[AND[AT E_i B_i : \rho 1][AT E_{i+1} B_{i+1} : \rho]] : \sigma$$

Definition 6 Probabilistic A-Box ($\mathbb{P}_{AB}$) in $\mathcal{BGTDL}$ for the whole scenario is the finite set of probabilistic axioms, $\mathbb{P}_A$, and role assertions set S_{RA}.

Definition 7 Probabilistic T-Box ($\mathbb{P}_{TB}$) in $\mathcal{BGTDL}$ for n different episodes is the finite set of n probabilistic basic patterns, i.e.,
$$\mathbb{P}_{TB} = \{\mathbb{P}_B 1, \mathbb{P}_B 2, \ldots, \mathbb{P}_B n\} \Rightarrow \mathbb{P}_{TB} = \{B_1 : \rho; B_2 : \rho; \ldots; B_n : \rho\}.$$

Definition 8 $\mathcal{BGTDL}$ knowledge base $\mathcal{BK} = (\mathbb{P}_{AB}, \mathbb{P}_{TB}, \mathcal{L})$.

Definition 9 Motion pattern ($\mathfrak{M}$) is of the form:
$\mathfrak{M} := CP : \Upsilon$ where CP is the composite pattern of the whole scenario, and Υ is its probability. CP is the conjunction of all n-1 probabilistic neighboring pattern, i.e.,

$$CP: \Upsilon \doteq [AND\ \mathbb{P}_{NP} 1, \mathbb{P}_{NP} 2, \ldots, \mathbb{P}_{NP} n\text{-}1] : \Upsilon \Rightarrow$$
$$CP: \Upsilon \doteq [AND\ NP_{1,2} : \sigma 1 ; NP_{2,3} : \sigma 2 ; \ldots; NP_{n-1,n} : \sigma n\text{-}1] : \Upsilon$$

Semantics The formal semantics of $\mathfrak{BGTDL}$ is the probabilistic distribution over the $\mathfrak{GTDL}$ interpretation.

1. **Annotated Interpretation**: An annotated interpretation is defined as $\mathfrak{I}{=}(\mathfrak{D},\mathcal{I})$ where $\mathfrak{D}$ is the domain and $\mathcal{I}$ the interpretation mapping that maps probabilistic axioms($\mathbb{P}_A$) to concept probability α. The annotated interpretation $\mathfrak{I}$ satisfies probabilistic axiom, $\mathsf{CA}^c : \eta$, denoted as $\mathfrak{I} \models \mathsf{CA}^c : \eta$ if $\mathfrak{I} \models \mathsf{CA}$, $\mathfrak{I}$ (c) $\in \mathfrak{D}$ and $\eta \in [0,1]$.

2. **Probabilistic Interpretation**: It is the probabilistic distributions over the probabilistic patterns,$\mathbb{P}$. Let us state this formally. A probabilistic interpretation $\mathfrak{P}$ is a probability function over the probabilistic patterns ($\mathbb{P}$) of a scenario that associates each Spatio_Temporal_Concept or spatial_Concepts with a positive probability. The probability μ of a probabilistic pattern,
   $\mathbb{P} : \mu$ in $\mathfrak{P}$, denoted as $\mathfrak{P}(\mathbb{P} : \mu)$, is computed using Logistic Function.

   (a) If $\mathbb{P}$ is of form: [AND [FILLS Primary_Object c][ALL $\mathsf{R_D}$ C_i]]:Ψ, then the conditional probability of node C_i (θ) from the conditional probability table of $\mathfrak{L}$ is taken as the weight. Logistic Function is then used to calculate the probability of $\mathbb{P}$, i.e., $\Psi = 1/(1+e^{-\theta})$.

   (b) If $\mathbb{P}$ is of the form: [AND $\{\mathfrak{STC}: \varphi\}^+$] :$\rho$ or [AND $\{\mathfrak{SC}: \varphi\}^+$] :$\rho$ where ρ is its probability , φ is taken as the weight of $\mathfrak{STC}$ or $\mathfrak{SC}$. Logistic Function then calculates the probability of each $\mathfrak{STC}$ or $\mathfrak{SC}$, i.e., $\mathfrak{P}_1 = 1/1+ e^{-\varphi}$. The joint probability is then found out to get ρ.

   (c) If the $\mathbb{P}$ is of the form [AND MEETS E_i E_{i+1} [AND [AT E_i B_i :$\rho 1$] [AT E_{i+1} B_{i+1} :$\rho 2$]] : σ where σ is its probability. Here, $\rho 1$ is taken as the weight of B_i and $\rho 2$ as the weight of B_{i+1}. Logistic Function then calculates the probability of each basic pattern, i.e., $\mathcal{P}\infty = 1/1+e^{-\rho}$. The joint probability of B_i is then found out.

The final motion pattern is of the form as in (b). Hence, the probability of final motion pattern $\mathfrak{M} \doteq \mathsf{CP}{:}\Upsilon$ in $\mathfrak{P}$ denoted as $\mathfrak{P}(\mathsf{CP}{:}\Upsilon)$ is obtained as in (b) such that $\Upsilon \in [0,1]$. An answer to the $\mathfrak{BGTDL}$ knowledge base ($\mathbb{P}_{AB}$, $\mathbb{P}_{TB}$, $\mathfrak{L}$) is $\mathsf{CP}{:}\Upsilon$

Semantic Properties $\mathfrak{BGTDL}$ faithfully extends both $\mathfrak{GTDL}$ and Logical Bayesian Network, $\mathfrak{L}$. Based on the semantics of $\mathfrak{BGTDL}$, it can be shown that there always exists all the three probabilistic patterns exhibited by a primary object during a scenario.

Following propositions show that the three probabilistic patterns ($\mathbb{P}$) can always be extracted from a given scenario.

Proposition 2 *Given a scenario and $\mathfrak{L}$, for an episode there always exists a probabilistic fundamental pattern per role and a single probabilistic basic pattern.*

Proof For an episode E_1 with roles $\mathsf{R_D}$, $\mathsf{R_O}$ and concepts Gi $\in \mathsf{C}_{D\,1}$, Gj $\in \mathsf{C}_{O\,1}$. Let us assume for an episode E_1 there exists two different probabilistic fundamental patterns ($\mathbb{P}_F$) per role. Thus, for role $\mathsf{R_O}$ we have,

$(\mathbb{P}_{\mathbb{F}|\nsucc}) = $ [AND [FILLS Primary_Object c][ALL $\mathsf{R_O}$ G1]]: $\xi 1$
$(\mathbb{P}_{\mathbb{F}\nsucc}) = $ [AND [FILLS Primary_Object c][ALL $\mathsf{R_O}$ G2]]: $\xi 2$
where G1, G2 $\in \mathsf{C_{O\,1}}$ and G1 and G2 are two different directions and c $\in$ Domain. But *episodes* as defined in Sect. 2.2 is a time period when an object does not change its qualitative spatial relation. Hence, G1 and G2 must be two different orientations of two different episodes. This contradicts the statement which defines an episode. Hence, our assumption that two probabilistic fundamental patterns exist for one role is wrong. Thus for role $\mathsf{R_O}$, there exists a single probabilistic fundamental pattern. Similar is the case for role $\mathsf{R_D}$. Thus, it follows that for an episode, there always exists a single probabilistic fundamental pattern per role.

A probabilistic basic pattern in terms of interpretation is as follows: $(\mathsf{P_B}) = \mathsf{B}_i : \rho = $ [AND $\mathsf{F_D}$ $\mathsf{F_O}$] : ρ. Since there exists one $(\mathbb{P}_{\mathbb{F}})$ per role, by Definition 4, an episode E_i with roles and concepts always comprises that there always exists a probabilistic basic pattern.   ∎

Proposition 3 *Given a scenario and $\mathfrak{L}$, for two neighbor episodes, there always exists a probabilistic neighboring pattern.*

Proof Consider two episodes E_1 and E_2 with roles $\mathsf{R_D}$, $\mathsf{R_O}$ and concepts $\mathsf{G}_i \in \mathsf{C_{D\,1}}$, $\mathsf{G}_j \in \mathsf{C_{O\,1}}$ such that E_1 and E_2 meets. Proposition 2 states that for an episode there always exists a probabilistic basic pattern. Thus for E_1 and E_2, their corresponding probabilistic basic patterns exist. Definition 5 defines that a probabilistic neighboring pattern consists of two probabilistic basic patterns of two episodes that meet each other. Hence, for two episodes that meet each other, there always exists a probabilistic neighboring pattern $\mathbb{P}_{\mathrm{NP}}$.   ∎

Once probabilistic basic patterns are obtained for a scenario with m episodes, probabilistic T-Box, $\mathbb{P}_{\mathrm{TB}}$ can be constructed. Proposition 4 is stated below.

Proposition 4 *Given a scenario and $\mathfrak{L}$, a probabilistic T-Box can always be constructed.*

Proof Consider a scenario S1. A scenario S1 always consists of m episodes, i.e., S1 = { $\mathsf{E}_1, \mathsf{E}_2, ..., \mathsf{E}_m$}. Definition 7 states that $\mathbb{P}_{\mathrm{TB}}$ is a finite set of m probabilistic basic patterns. Proposition 2 depicts that for an episode there always exists a single probabilistic basic pattern. Thus for m episodes, m $\mathsf{P_B}$ exists. This m $\mathsf{P_B}$ encodes the $\mathbb{P}_{\mathrm{TB}}$. Hence, given a scenario and $\mathfrak{L}$, a probabilistic T-Box can always be constructed.   ∎

Proposition 5 *Given a scenario with roles, concepts, and m episodes $\mathsf{E}_1, \mathsf{E}_2, ...,$ E_m such that each pair of succeeding episodes is neighbors to each other: We can always find a motion pattern for that scenario.*

Proof Consider a scenario S1 with m episodes, i.e., S1 = { $\mathsf{E}_1, \mathsf{E}_2, ..., \mathsf{E}_n$ }. Definition 9 states that a motion pattern comprises probabilistic neighboring patterns. Proposition 3 shows that there always exists a probabilistic neighboring pattern for two episodes that meet. Thus for m different episodes, m-1 probabilistic neighboring pattern $(\mathbb{P}_{\mathrm{NP}})$ exists. Once $\mathbb{P}_{\mathrm{NP}}$ are obtained, by the semantics of $\mathfrak{GIDL}$, motion pattern can be always found out for a given scenario.   ∎

4 Motion Patterns Recognition in an Uncertain Domain

Motion pattern recognition using $\mathcal{BGTDL}$ can be done using *entailment*. Entailment involves making conclusions based on entailed information of the knowledge base. We discuss below motion pattern recognition as entailment proof. The example from [3] is extended to uncertain domains.

Illustrative Example: Example scenario in Fig. 1 is an uncertain domain. Three episodes E_1, E_2, and E_3 can be demarcated. A is the primary object, and B is the reference object. Following are the orientation and direction relations of A w.r.t B:

1. $E_1 \rightarrow$ Spatial orientation : Back & Right, Direction : Same.
2. $E_2 \rightarrow$ Spatial orientation : Right, Direction : Same.
3. $E_3 \rightarrow$ Spatial orientation : Front&Right, Direction : Same. Roles: Primary object, R_O, R_D.

Further, we could identify the following relationship between the episodes.

$$P1 \doteq [MEETSE_1E_2]//E_1 \quad \text{and} \quad E_2 \text{ meets.} \tag{1}$$

$$P2 \doteq [MEETSE_2E_3]//E_2 \quad \text{and} \quad E_3 \text{ meets.} \tag{2}$$

$C_{D1} = \{Same\}$; $C_{O1} = \{Back\&Right, Right, Front\&Right\}$.
$\mathcal{L}$ for both the roles R_D and R_O are constructed as shown in Fig. 2.
True concepts: Back&Right, Right, Front&Right, Same

$$\alpha(Back\&Right) = 1/3 = 0.33; \ \alpha(Right) = 1/3 = 0.33$$
$$\alpha(Front\&Right) = 1/3 = 0.33; \ \alpha(Same) = 3/3 = 1.$$

$\mathbb{P}_\mathbb{A} = \{(Same(A), Back\&Right(A):0.33), (Same(A), Right(A):0.33),$
$(Same(A), Front(A):0.33)\}$
The S_{RA} and P_A yield the probabilistic A-Box. During E_1, A is in the Back&Right

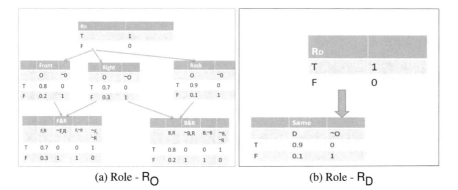

(a) Role - R_O (b) Role - R_D

Fig. 2 Logical Bayesian Network

of B and is in the same direction as that of B with probability 0.446. During E_2, A is in the Right of B along the same direction with probability 0.443. A is in the Front of B along the same direction with probability 0.443 during E_3.

$P_{B1} \doteq$ [AND[AND[FILLS Primary Object A][ALL R_D Same]] :0.709

[AND[FILLS Primary Object A][ALL R_O Back&Right]]:0.690]:0.446

$P_{B2} \doteq$ [AND[AND[FILLS Primary Object A][ALL R_D Same]] :0.709

[AND[FILLS Primary Object A][ALL R_O Right]]:0.667]:0.443

$P_{B3} \doteq$ [AND[AND[FILLS Primary Object A][ALL R_D Same]] :0.709

[AND[FILLS Primary Object A][ALL R_O Front]]:0.667]:0.443

$Q1 \doteq [ATE_1 P_{B1}] : 0.446 //$A is exhibiting B_1 during E_1 with probability 0.446

$$(3)$$

$Q2 \doteq [ATE_2 P_{B2}] : 0.443 //$A is exhibiting B_2 during E_2 with probability 0.443

$$(4)$$

$Q3 \doteq [ATE_3 P_{B3}] : 0.443 //$A is exhibiting B_3 during E_3 with probability 0.443

$$(5)$$

$$\mathbb{P}_{TB} = \{P_{B1}, P_{B2}, P_{B3}\} \qquad (6)$$

Proof Sketch From (6)

$\mathfrak{BK} \models P_{B1}.$

$P_{B2}.$

$P_{B3}.$

Also, from (1) and (2),

$\mathfrak{BK} \models P1.$

$P2.$

From Eq. (3), (4) and (5)

$\mathfrak{BK} \models Q1.$

$Q2.$

$Q3.$ ∎

Above leads to **Assertion** 1 and **Assertion** 2. These in turn lead to inference on the motion pattern of A w.r.t B.

Assertion 1 *E_1 and E_2 are neighbor episodes. A and B are moving in the same direction, and A moves from back right of B to the right of B.*

$\mathfrak{BK} \models \mathbb{P}_{NP}1 \doteq$ [AND[MEETS E_1 E_2][AND[AT E_1 $\mathbb{P}_{B\cancel{K}}$][AT E_2 $\mathbb{P}_{B\cancel{K}}$]]]:0.371

Assertion 2 *E_2 and E_3 are neighbor episodes. A and B are moving in the same direction, and A moves from right of B to front of B.*

$\mathfrak{BK} \models \mathbb{P}_{NP}2 \doteq$ [AND[MEETS E_2 E_3][AND[AT E_2 $\mathbb{P}_{B\cancel{K}}$][AT E_3 $\mathbb{P}_{B\cancel{K}}$]]]:0.3709.

Inference 1 *A moves from the back right of B to the front while moving in the same direction as that of B.*

$\mathfrak{BK} \models MP \doteq$ [AND $\mathbb{P}_{NP}1$ $\mathbb{P}_{NP}2$]:0.137

5   Conclusion

Unification of QSTR, $\mathcal{GTDL}$, and $\mathcal{L}$ present a formal basis for tenable reasoning. The use of $\mathcal{BGTDL}$ reinforces the framework supporting spatiotemporal reasoning in an uncertain domain. There are innumerable ways in which the framework proposed can be extended. Future work includes validation of our framework with more datasets that would include richer context sources.

References

1. S.M. Hazarika, *Qualitative Spatial Change: Space-time Histories and Continuity*. Ph.D. Thesis, University of Leeds, School of Computing, 2005
2. A.G. Cohn, S.M. Hazarika, Qualitative spatial representation and reasoning: an overview. Fundam. Inf. **46**(1–2), 1–29 (2001)
3. U. Talukdar, R. Baruah, S.M. Hazarika, A description logic based QSTR framework for recognizing motion patterns from spatio-temporal data, in *2nd International Conference on Recent Trends in Information Systems (ReTIS)* (IEEE, New York, 2015), pp. 38–43
4. R. Baruah, S.M. Hazarika, Reasoning about directions in an egocentric spatial reference frame, in *Proceedings of DLAC 2013* (2013)
5. R. Baruah, *Representation and Recognition of Motion Patterns Using a Qualitative Language Based Framework*. PhD Thesis, Tezpur University, School of Engineering, 2014
6. R. Baruah, S.M. Hazarika, Qualitative directions in egocentric and allocentric spatial reference frames. Int. J. Comput. Inf. Syst. Ind. Manage. Appl. **6**, 344–354 (2014)
7. S. Dodge, R. Weibel, A.K. Lautenschtz, Towards a taxonomy of movement patterns. Inf. Visual. **7**, 240–252 (2008)
8. K.S. Dubba, A.G. Cohn, D.C. Hogg, Event model learning from complex videos using ILP, in *ECAI 2010* (2010), pp. 81–103
9. A.G. Cohn, J. Renz, M. Sridhar, Thinking inside the box: a comprehensive spatial representation for video analysis, in *Principles of Knowledge Representation and Reasoning: Proceedings of the Thirteenth International Conference, KR 2012* (AAAI Press, 2012)
10. R.J. Brachman, H.J. Levesque, Knowledge representation and reasoning (Chap. 9), in *Structured Descriptions*(2004), pp. 155–180
11. D. Koller, A. Levy, A. Pfeffer , P-CLASSIC: a tractable probabilistic description logic, in *Proceedings of the 14th National Conference on Artificial Intelligence (AAAI-97)* (1997), pp. 390-397
12. A. Schmiedel, *A Temporal Terminological Logic*. Technical University, Projektgruppe KIT (1990), pp. 640–645
13. B. Hummel, W. Thiemann, I. Lulcheva, Scene understanding of urban road intersections with description logic, in *Logic and Probability for Scene Interpretation* (2008), 8091
14. X. Wang, L. Chang, Z. Li, Z. Shi, A dynamic description logic based system for video event detection. Front. Electr. Electron. Eng. China **5**(2), 137–142 (2010)
15. M. Sridhar, *Unsupervised Learning of Event and Object Classes from Video*. Ph.D. Thesis, University of Leeds, School of Computing, 2010

Formants and Prosody-Based Automatic Tonal and Non-tonal Language Classification of North East Indian Languages

Chuya China Bhanja, Mohammad Azharuddin Laskar and Rabul Hussain Laskar

Abstract This paper proposes an automatic tonal and non-tonal language classification system for North East (NE) Indian languages using formants and prosodic features. The state-of-the-art system for tonal/non-tonal classification uses mostly prosodic features and considers the utterance-level analysis unit during feature extraction. To this end, the present work explores formants and studies if it has complimentary information with respect to prosody. It also analyzes different analysis units for feature extraction, namely syllable, di-syllable, word, and utterance. Classification techniques based on Gaussian mixture model—universal background model (GMM-UBM), neural network and i-vector have been explored in this work. The paper presents NIT Silchar language database (NITS-LD) prepared in-house to carry out experimental validation. It covers seven NE Indian languages and uses data from All India radiobroadcast news archives. Experimental analysis suggests that artificial neural network (ANN) based on syllable level features provides the lowest EERs of 31.8, 36 and 37.8% for test data of durations, 30, 10, and 3 s, respectively, when the combination of prosodic features and formants are used. The addition of formants helps to improve the system performance by up to 6.8, 7.8 and 9.2% for test data of the three different durations with respect to that of prosodic features.

Keywords Tonal and non-tonal languages · Formants · Prosody · Syllables · Classifiers · Database

1 Introduction

The aim of automatic language identification (LID) systems is to detect the language identity of an uttered speech. LID system's performance is largely dependent on the count of target languages. In the case where target language count is large, it is considered useful to pre-classify languages into distinct groups or different language

C. C. Bhanja (✉) · M. A. Laskar · R. H. Laskar
Department of Electronics and Communication Engineering, NIT
Silchar, Silchar 788010, Assam, India
e-mail: chuya.bhanja@gmail.com

© Springer Nature Singapore Pte Ltd. 2020
A. Elçi et al. (eds.), *Smart Computing Paradigms: New Progresses and Challenges*,
Advances in Intelligent Systems and Computing 767,
https://doi.org/10.1007/978-981-13-9680-9_14

families. In one study, Wang et al. [1] reported a novel system for tonal/non-tonal classification of six world languages (CALLFRIEND Mandarin Chinese-Mainland Dialect, CALLFRIEND Mandarin Chinese-Taiwan Dialect, CALLFRIEND American English-Southern Dialect, CALLFRIEND American English-Non-Southern Dialect, CALLFRIERD Vietnamese and CALLFRIEND German) using utterance-level pitch and duration features in conjunction with ANN classifier. Their system could achieve an accuracy of 80.6% having used phonetically labeled data. The need for phonetically labeled data puts a constraint on the system in situations where a linguistic expert or phonetically labeled data is unavailable. The present work, therefore, aims to prepare a tonal/non-tonal classification system, which is not dependent on phonetically labeled data or any automatic speech recognizer (ASR).

A study revealed that almost half of the world languages are tonal, at least, to some moderate extent [2], and there is a fine difference between these two language categories. For tonal languages, pitch-changes within a syllable follow a regular pattern. In addition, the tone is effectively correlated with prosody, such as energy profile and duration [3]. This paper, therefore, explores effective parameterization of prosodic features that may help increase its discriminating ability.

Though prosody is an important parameter for tonal/non-tonal classification, only prosody-based LID system performs poorly when compared to state-of-the-art LID system based on spectral features. Formants can be used as an important feature for the discrimination of tonal languages from non-tonal. The resonance frequencies of the vocal tract can be approximated as the formants information of the speech [4]. For tonal languages like Mandarin Chinese, some researchers have concluded that vowel information has a perceptual advantage over tonal information [5]. Since the vowel sounds resonate in the throat, formants turn out to be useful in determining vowel sounds. Human ear and brain utilize this useful information of speech, which is also robust to wideband noise, to distinguish between sounds [6]. Vowel sounds are uttered differently in different languages, and hence formants can be used as vital information for discriminating among different languages. Thus, formants features are investigated in the context of the classification problem addressed in this work. In addition, this paper further goes on to demonstrate the effect of combining formants and prosodic information on the performance of this classification task.

Literature confirms the alignment of tonal events with segmental events for tonal languages [7]. In tonal languages, the pitch contour peaks and valleys are found to align with the onsets and offsets of syllables. Most of the languages from Northeastern India are syllable centric [8] and the syllables itself possess the evidence of language-specific cues. Thus, the present work proposes to use syllabic level analysis of the speech signal for tonal/non-tonal classification.

This work focuses mainly on the classification of NE Indian languages, which are known to be closely related. Language diversity is an interesting phenomenon in the Northeastern region of India. Here, the influence of one language on the other is comparatively high. This makes it more difficult to discriminate among these languages. A pre-classification module, in such a case, can be very helpful in improving the accuracy of the LID system. The available language resources do not cover NE Indian languages to that extent. In an effort to overcome this limitation,

NITS-LD has been prepared. It covers seven languages—Manipuri, Assamese, Mizo, Bengali, Indian English, Nagamese, and Hindi. A total of 4 h of data per language has been collected from All India Radio news archives to prepare the database. This paper thus deals with the development of tonal and non-tonal language classification system for NE Indian languages.

Three different modeling techniques, namely i-vector-based SVM, GMM-UBM, and ANN have been explored in this work. A comparative analysis of all the three modeling techniques in the context of tonal/non-tonal classification of languages is presented in this paper.

The contributions of this work may be summarized as follows:

- This work attempts to develop an automatic tonal/non-tonal language classification system that may not be dependent on any phonetic information.
- Formants have been introduced as spectral features for the tonal/non-tonal classification task.
- As the tonal characteristics reflect on the onset and offset of a syllable, this work proposes to consider syllables as basic units.
- NITS-LD, covering seven NE Indian languages has been presented in this paper. In addition, the classification systems have been analyzed in this database.
- The work explores three different modeling techniques, namely i-vector-based SVM, GMM-UBM, and ANN, in the context of tonal/non-tonal classification problem.

The paper has been structured as follows: Detailed description of the proposed system is given in Sect. 2, while Sect. 3 gives a description of the databases considered in this work. Features and classifiers have been described in Sects. 4 and 5, respectively. Experimental results and discussions are given in Sects. 6 and 7, finally, concludes the work and states the future scope of work.

2 Proposed System for Language Classification

Here, a system has been prepared for tonal/non-tonal language classification at syllabic level in an attempt to improve the accuracy with respect to the state-of-art system. Instead of using utterance as a whole, syllables have been used for feature extraction in this case. Every language has only a finite set of syllables. There exist certain common structures in syllable formation, such as vowel (V), vowel–consonant (VC), vowel–consonant–consonant (VCC), consonant–vowel–consonant–consonant (CVCC), etc. The syllables in Indian languages are mostly of CV type [9]. Every spoken utterance can be treated as a sequence of syllables having a rhythmic property which is associated with the opening and the closing of the mouth at the time of speaking [9]. Tonal characteristics are sometimes aligned with the syllable onset and offset points [9]. Also, pitch changes in a regular manner within a syllable for tonal languages. This gives the motivation to look for some language-specific cues at the syllable level. In order to obtain syllable-based prosody and formants,

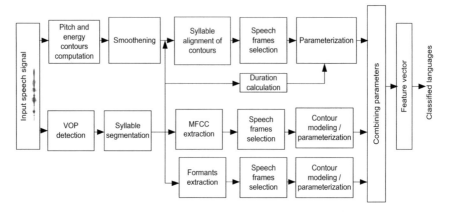

Fig. 1 Schematic diagram of the proposed language classification system

firstly, the CV regions are identified directly from the speech signal. Syllables are obtained using the concept of vowel onset point (VOP) [10]. Figure 1 depicts the schematic diagram of the proposed language classification system.

3 Database Used for Language Classification

The OGI-MLTS speech [11] corpus is a corpus of spontaneously spoken, fixed-vocabulary speech. It covers 11 languages—English, Hindi, Korean, Farsi, German, French, Japanese, Mandarin Chinese, Vietnamese, Spanish, and Tamil. Data have been collected over a telephone line at a sampling frequency of 8 kHz. Mandarin and Vietnamese are the two tonal languages in this database, while the remaining nine are non-tonal. For experimentations, 10 languages of this database (leaving out Japanese) have been considered in this work. This database, however, does not include any NE Indian languages. The NITS language database helps to study the NE Indian languages in particular. It comprises of seven North East Indian languages, namely Manipuri, Assamese, Mizo, Bengali, Indian English, Nagamese, and Hindi. The data have been collected from AIR news archives. Of these languages, Mizo and Manipuri are tonal in nature and the remaining are non-tonal. The particulars of the two databases are tabulated in Table 1.

4 Features Used in Language Classification

In this work, a collection of prosodic features, namely energy contour, pitch contour, and duration, and spectral features, namely the first four formants are used as front-

Table 1 Comparison of databases, OGI-MLTS and NITS-LD

Features	OGI-MLTS database	NITS-LD
Sampling frequency	8 kHz	8 kHz
No. of target languages	11	7
No. of speakers	90 per language	Assamese—50, Bengali—40, English—21, Hindi—20, Manipuri—13, Mizo—8, and Nagamese—6
Mode of speech	Spontaneous	Scripted
Channel conditions	Noisy	Non-noisy
Environment	Realistic	Studio
Channel characteristics	Different	Similar

end features. Extraction and parameterization of different features are discussed in the following sections.

4.1 Extraction of Prosodic Features

The first steps toward the calculation of prosodic features involve the computation of pitch and energy contours of the speech signal. Pitch is calculated based on robust algorithm for pitch tracking (RAPT) algorithm [12]. To obtain the energy contour, energy values are computed for every 10 ms speech frame. Both pitch and energy contours are processed with fifth order median filter for smoothening. Thereafter, the vowel onset points (VOPs) [10] are aligned with both the contours. The speech units between every two consecutive VOPs are considered as syllables. These contours corresponding to these syllables are then parameterized in terms of linear combination of Legendre polynomials [6]. The non-speech frames are detected using RAPT algorithm and are discarded. The syllable units whose length is less than 50 ms are not used. The frame count in a syllable is considered as the duration feature. Duration is further parameterized to obtain the rhythm feature. The rhythm of a syllable is defined as the ratio of the duration of the voiced region of the syllable to that of the whole syllable.

4.2 Extraction of Formants

Formants are used as the spectral feature in this experiment. Vocal tract, in general, is considered as a time-invariant, all-pole filter and each of the conjugate pair of poles where resonance occurs correspond to a formant frequency. In this analysis, formants are computed using 12th order linear predictive (LP) analysis. Speech is framed into blocks of 20 ms with 50% overlapping and multiplied with DFT Hamming window. LP coefficients are computed for every 20 ms frame of the syllable. Formant frequencies are calculated from these LP coefficients, of which the first four (F_1, F_2, F_3, and F_4) are used in this classification task. Contour corresponding to each formant of a syllable is again parameterized with a linear combination of Legendre polynomials [6] using Eq. (1).

4.3 Parameterization of Different Features

The mathematical representation of the parameterization process using Legendre polynomials is shown in Eq. (1) [6].

$$f(t) = \sum_{n=0}^{N} a_n p_n(t) \tag{1}$$

where $f(t)$ denotes the contour, a_n is nth coefficient of Legendre fit, and $p_n(t)$ is the nth Legendre polynomial. Each of the coefficients of Legendre fit represents different parameters of the contour, like mean, slope, curvature, etc. Higher order coefficients give the finer details of the contour. Fourth order Legendre polynomials are used, resulting in 5-dimensional pitch, 5-dimensional energy, and 20-dimensional formant features (each formant has 5 dimensions) for a syllable. Besides, 2-dimensional (syllable duration and rhythm) duration feature is also considered. The prosody and formants parameters are all stacked in a row to form the final feature vector.

4.4 Data Normalization

To minimize the effect of different kinds of variabilities, such as speaker and channel variabilities, the data need to be normalized before applying machine-learning algorithms. In this study, the data have been normalized using the z-normalization technique when using i-vector and GMM-UBM-based classifiers. For ANN classifier, the features are normalized to bring the values to lie in the range between -1 and $+1$.

5 Classifiers Used in Language Classification

Classification of languages can be done using a generative approach, a discriminative approach, or a combination of generative and discriminative approaches. The classification in this paper follows both generative and discriminative approaches. In the present study, GMM-UBM [13], i-vector-based SVM [15], and ANN [14] are used to develop the language classification systems. In the case of the i-vector system, within-class covariance normalization (WCCN) [16] is used for data normalization.

6 Experimental Results

6.1 Experimental Setup

NITS-LD and OGI-MLTS databases have been used for experimental validation. Around 2 h of speech, data of each of the languages are used in the training set. The performance of the language classification system is highly dependent on the duration of the test data. Therefore, different test data durations, namely 30, 10, and 3 s have been experimented with in this work. In all, 200 test utterances—100 from tonal and another 100 from non-tonal languages—from each of the databases have been used for evaluation. It is ensured that the training and the testing data are mutually exclusive.

Here, prosodic features are denoted by A_1, prosodic features $+F_1$ by A_2, prosodic features $+ F_1 + F_2$ by A_3, prosodic features $+ F_1 + F_2 + F_3$ by A_4, and prosodic features $+ F_1 + F_2 + F_3 + F_4$ by A_5. In this experiment, an equal error rate (EER) is used as a measure of performance evaluation. The scores from all the units of the test utterance are averaged to obtain an utterance-level score based on which decision is made.

6.2 Experimental Results

6.2.1 Syllable-Level Performance Analysis

This section analyzes the performance of different classifiers using different feature sets, A_1–A_5, extracted from syllabic units. Performance of a GMM-UBM model depends heavily on the number of Gaussian components considered in the modeling. Experiments have, therefore, been conducted varying the number of Gaussian components, particularly 2, 4, 8, 16, 32, 64, 128, etc., to analyze this aspect. Different features report their best results for a different number of Gaussian components—for A_1, 16; for A_2, 128; for A_3, 128; for A_4, 256; and for A_5, 256. Table 2 presents the best results for the GMM-UBM model for features A_1–A_5.

Similarly, when experimenting with ANN, different network structures have been explored. It is observed that 12L-20N-1L, for A_1; 17L-29N-8N-1L, for A_2; 22L-35N-10N-1L, for A_3; 27L-39N-12N-1L, for A_4; and 32L-45N-20N-1L, for A_5 provide the best individual results in case of NITS-LD. The upper limit on epochs is set at 500 and the activation function used is Tan-sigmoid. Table 3 shows the EER values obtained

Table 2 Performance of GMM-UBM classifier for NITS-LD

Duration of test data (s)	EER (%)				
	A_1	A_2	A_3	A_4	A_5
30	40	38.1	33	33.8	34
10	44	41	36.1	36.6	38.6
3	48	44.8	39	40.1	42

Table 3 Performance of ANN classifier for NITS-LD

Duration of test data (s)	EER (%)				
	A_1	A_2	A_3	A_4	A_5
30	38.6	36	31.8	32.6	33
10	43.8	40.8	36	36.8	37.4
3	47	43.6	37.8	40.2	41

for different features when using ANN as the classifier. The final score is obtained by averaging the syllable-level scores across the whole utterance.

From experimental results, it is observed that for A_1, 16 Gaussian components and 100-dimensional total variability (TV) matrix; for A_2, 64 Gaussian mixtures and 200 dimensional TV matrix; for A_3, 128 Gaussian mixtures and 200-dimensional TV matrix; for A_4, 256 Gaussian mixtures and 300-dimensional TV matrix; for A_5, 256 Gaussian mixtures and 300-dimensional TV matrix provide the lowest EERs in their respective cases. Here, linear kernel is used in SVM and the scores are first converted into posterior probabilities with the help of optimal sigmoid transformation. For final score calculation, again, the syllable-level scores are averaged across the utterance. Table 4 shows the accuracies of i-vector-based SVM classifier.

Experimental results for the different classifiers on OGI-MLTS database using features, A_1–A_5, are given in Table 5. Modeling parameters of all the three classifiers are the same as that used in case of NITS-LD.

The following observations can be made from Tables 2, 3, 4, and 5:

- Out of the first four formants, only the first two formants are significant. Also, from experimental results, it can be observed that between these two formants, the second formant is more significant than the first.

Table 4 Performance of i-vector based SVM classifier for NITS-LD

Duration of test data (s)	EER (%)				
	A_1	A_2	A_3	A_4	A_5
30	40.8	38.6	34	34.6	35
10	46.7	43.8	38.4	39.7	40.8
3	49.8	46	41.2	42	42.6

Table 5 Performance of different classifiers on OGI-MLTS database

	EER (%)				
	A_1	A_2	A_3	A_4	A_5
GMM-UBM					
30 s	43.2	42	38.5	39.1	40
10 s	49.5	48.4	45	45.9	46.8
3 s	53.8	50.6	46.8	47.4	48.3
ANN					
30 s	38	36.4	33.3	33.8	34.6
10 s	41.2	39.7	37	38	38.4
3 s	49.6	48	46.2	48	49.5
i-vector-based SVM					
30 s	43.7	41.7	38	40.2	41
10 s	50.6	49	45.8	46	47.3
3 s	54	52.8	48.6	49.2	50.6

- Combination of prosodic features and formants results in the lowest EER.
- 30 s duration test data provide the lowest EER. It can be inferred that the system captures more variation from 30 s duration of test data than from 10 and 3 s test data.
- ANN is found to give the best performance. i-vector-based SVM classifier gives the next best results. It may, therefore, be inferred that compared to the i-vector and GMM-UBM based models, ANN is better suited to learn and code language-specific information that can help discriminate among languages.

In general, it is observed that the different classifiers provide better or at least comparable performance for NITS-LD over OGI-MLTS database. This may be attributed to two factors—(a) target languages are lesser in number in NITS-LD than that in OGI-MLTS database and (b) NITS-LD has noise-free data, while OGI-MLTS database includes noisy data.

6.2.2 Performance Comparison of Features Extracted of Different Analysis Units

System performance has been analyzed for different analysis units, viz. di-syllables, words, and utterance as a whole using A_3 features. Utterance-level features are extracted from the whole utterance. For di-syllable units, feature parameters of the two consecutive syllables are concatenated, while for word units, the features of the preceding and the succeeding syllables are concatenated with that of the present syllable. From Table 6, it can be observed that syllabic level features result in the best performance. It may, therefore, be inferred that syllable-level analysis is best suited for this task.

Table 6 Comparative performance analysis of different analysis units on NITS-LD

Analysis level	EER (%) for 30 s test data		
	GMM-UBM	ANN	i-vector-based SVM
Utterance	40	38.6	44
Syllable	33	31.4	34
Di-syllable	37.3	36.4	40.1
Word	35.4	35.8	37.6

7 Conclusion

This paper presents a classification system for discriminating tonal languages from non-tonal without depending on any phonetic knowledge. The importance of prosodic features and their combination with spectral features for tonal/non-tonal classification of languages has been discussed in this paper. This paper also analyzes how the system performance changes for the sequential addition of the first four formants with prosodic features. Ten languages from OGI-MLTS database and seven more from NITS-LD have been considered for experimental validation of the systems. It explores different analysis units for feature extraction. Experimental results suggest that syllables are best-suited analysis units for this classification problem. Three different test data durations have been tested for three different classifiers. It is observed that 30 s test data provides the lowest EER and ANN classifier proves to be the best classifier in terms of accuracy, both for NITS-LD and OGI-MLTS. In the future, the system performance can be analyzed further for noisy data. In addition, some new features may be explored to improve the system performance.

Acknowledgements The authors acknowledge TEQIP III (NIT Silchar) for funding participation in the conference.

References

1. L. Wang, E.E. Ambikairajah, H.C. Choi, Automatic tonal and non-tonal language classification and language identification using prosodic information, in *International Symposium on Chinese Spoken Language Processing, (ISCSLP)* (2006), pp. 485–496
2. D. Dan, D. Robert Ladd, Linguistic tone is related to the population frequency of the adaptive haplogroups of two brain size genes, ASPM and microcephalin, in *PANS* (2007). https://doi.org/10.1073/pnas.0610848104
3. C. Qu, H. Goad, The interaction of stress and tone in standard Chinese: Experimental findings and theoretical consequences, in *Tone: Theory and Practice, Max Planck Institute for Evolutionary Anthropology* (2012)
4. B. Gold, L. Rabiner, Analysis of digital and analog formant synthesizers. IEEE Trans. Audio Electroacoust. **16**, 81–94 (1968)
5. H.-N. Lin, C.-J.C. Lin, Perceiving vowels and tones in Mandarin: The effect of literary Phonetic systems on phonological awareness, in *Proceedings of the 22nd North American Conference on*

Chinese Linguistics (NACCL-22) and *The 18th International Conference on Chinese Linguistics (ICCL-18)*, Harvard University, Cambridge, 2010, pp. 429–437

6. D. Martinez, E. Lleida, A. Ortega, A. Miguel, Prosodic features and formant modelling for an I-vector based language recognition system, in *ICASSP* (2013), pp. 6847–6851
7. M. Atterer, D.R. Ladd, On the phonetics and phonology of "segmental anchoring" of F0. J. Phon. **32**, 177–197 (2004)
8. A.K. Singh, A computational phonetic model for Indian language scripts, in *Constraints on Spelling Changes: Fifth International Workshop on Writing Systems*, Nijmegen, The Netherlands (2006)
9. L. Mary, B. Yegnanarayana, Extraction and representation of prosodic features for language and speaker recognition. Speech Commun. **50**(10), 782–796 (2008)
10. S.R.M. Prasanna, B.V.S. Reddy, P. Krishnamurthy, Vowel onset point detection using source, spectral peaks, and modulation spectrum energies. IEEE Trans. Audio Speech Lang. Process. **17**, 556–565 (2009)
11. Y. Muthusamy, R. Cole, B. Oshika, The OGI multi-language telephone speech corpuses, in *Proceedings of International Conference Spoken Language Processing (ICSLP)* (1992), pp. 895–898
12. D. Talkin, A robust algorithm for pitch tracking (RAPT), in *Speech Coding and Synthesis*, ed. by W.B. Klein, K.K. Paliwal (Elsevier, New York, 1995)
13. D. Reynolds, *Gaussian Mixture Models. Encyclopedia of Biometric Recognition* (Springer, New York, 2008)
14. B. Yegnanarayana, *Artificial Neural Networks* (Prentice-Hall of India Private Limited, New Delhi, 2005)
15. N. Dehak, P. Torres-Carrasquillo, D. Reynolds, R. Dehak, Language recognition via I-vectors and dimensionality reduction, in *Interspeech Conference*, Florence, Italy (2011), pp. 857–860
16. A.O. Hatch, S. Kajarekar, A. Stolcke, Within-class covariance normalization for SVM-based speaker recognition, in *Proceedings ICSLP* (2006), pp. 1471–1474

Operating System, Databases, and Software Analysis

On Ensuring Correctness of Cold Scheduler

Aparna Barik, Nabamita Deb and Santosh Biswas

Abstract Cold scheduling is one widely used transformation technique that reorders the instruction sequence in such a way bit switching between two successive instructions will be minimum and thus reduces power consumption. To make the system reliable, ensuring correctness of cold scheduling is important. In this paper, a method for verification of cold scheduler is presented. Our method first extracts finite state machines with data paths (FSMDs) from the input and the output of a cold scheduler and then applies an FSMD-based equivalence checking method to ensure the correctness of the cold scheduler. We have implemented our method and tested on some examples. The results show the effectiveness of our method.

Keywords Cold scheduling · Verification · Equivalence checking · FSMD

1 Introduction

The reduction of power consumption becomes an important design consideration together with performance in embedded systems. In compiler optimization, by applying power reduction technique, a new version of program can be obtained which consumes low power [9]. The authors in [3] explained reducing bit switching is one of the techniques to reduce power consumption. A bit switch from 0 to 1 and 1 to 0 consumes more energy than switch from 0 to 0 or from 1 to 1. When a processor moves to the current instruction from the previous instruction, some bit switching happens. So, the energy consumed by an instruction depends partially on that previous instruction. During instruction scheduling, scheduler can rearrange instructions without violating data dependencies. The cold scheduling [7] is one of the techniques that tries to reduce bit switching. Basically, cold scheduling uses traditional performance-driven

A. Barik (✉) · N. Deb
Gauhati University, Gauhati, Assam, India
e-mail: barik.aparna@gmail.com

S. Biswas
IIT Guwahati, Guwahati, Assam, India

© Springer Nature Singapore Pte Ltd. 2020
A. Elçi et al. (eds.), *Smart Computing Paradigms: New Progresses and Challenges*,
Advances in Intelligent Systems and Computing 767,
https://doi.org/10.1007/978-981-13-9680-9_15

scheduling technique which schedules instruction in such a way that bit switching between successive instructions is low. Reducing bit switching through instruction scheduling significantly improves the power consumption in embedded system. Cold scheduling is applied in assembly-level code.

Modern-day software is complex and consists of thousands of lines of code. Thus, any software of this size is prone to have implementation errors. Verification of the programs/software process catches bugs which cannot be uncovered during testing and even during long-term use of it. The objective of this work is to ensure correctness of the cold scheduling optimization. Various researchers target verification of loop transformations, code motions, copy propagation, constant propagation, dead code elimination, thread-level parallelizing transformation, etc [5, 8]. To the best of our knowledge, there is no approach yet to verify cold scheduling optimization.

In this work, we have proposed a verification framework to ensuring correctness of cold scheduler. The overall flow of our tool is given in Fig. 1. As shown, we will extract FSMDs from both the input and the output assembly programs of the cold scheduler. We then apply our FSMD-based equivalence checker to check the equivalence between the input and the output FSMDs. We have also created an FSMD to dotty format parser to visualize the input and output FSMDs using Graphviz software [1]. This will help to debug the non-equivalence. The working of our method is explained with an example. The proposed method is implemented and tested with test cases. Specifically, the contributions of this work are as follows:

We have organized the paper as follows. In Sect. 2, the cold scheduling is explained with an example. The FSMD-based equivalence checking method is presented in 3. The FSMD extraction from assembly code is explained in Sect. 4. The verification of cold scheduling is explained with example in Sect. 5. The experimental results are explained in Sect. 6. The paper is concluded in Sect. 7.

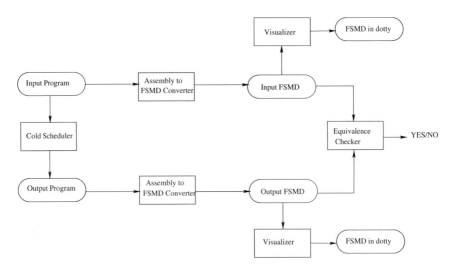

Fig. 1 Verification of cold scheduling

2   Cold Scheduling

The cold scheduling is illustrated with an example given in Table 1. This example is adapted from [3]. In this example, two-address register machine is considered and the opcodes are three bits wide. For example, we are considering opcodes 1 (=001) to 6 (=110) in the same order they are in the table. There are four registers in the machine, numbered 0 to 3. So, the register field is two bits wide. For example, we will check the difference in the bits of two consecutive operations. The assembly code given in Table 1 is a set of instructions of the code for the expression $b*b - 4*(a*c)$. The column 1 of the table is showing the initial code, and the fourth column is showing the code after cold scheduling. From the above code sequences, we can see that input code generates 20 bit switches. The output generated by cold scheduler has 9 bit switches. Therefore, the cold scheduler output improves the bit switching by 55%.

For this example, there can be 168 possible schedules (as topological sorts obeying the dependencies). Also, there can be 4! means 24 possible register assignments per schedule. So, there can be total 168 * 24 = 4032 possible code sequences. Obviously, not all of the 4032 possibilities satisfy the data dependency of the initial program. The cold scheduler should generate only the correct sequence. The objective of this work is *to ensure that the output sequence generated by cold scheduler is satisfying the data dependence of the input program.* In other words, cold scheduler does not violate the data dependence of the input programs.

3   FSMD-Based Equivalence Checking

Following definition of the FSMD model has been taken from [2] which is proposed by Gajski et al. The equivalence checking method is adapted from [5].

Table 1 An example of cold scheduling

Instructions	Opcode-reg	Bit switch	Instructions	Opcode-reg	Bit switch
mov b, R0	010..00	0	*mov* a, R1	010..01	0
mul R0, R0	101..00	3	*mov* b, R0	010..00	1
mov a, R1	010..01	4	*mov* c, R2	010..10	1
mov c, R2	010..10	2	*mov* 4, R3	001..11	3
mul R2, R1	101..01	5	*mul* R2, R1	101..01	2
mov 4, R3	001..11	2	*mul* R3, R1	101..01	0
mul R3, R1	101..01	2	*mul* R0, R0	101..00	1
sub R1, R0	100..00	2	*sub* R1, R0	100..00	1
(a) straightforward		20	(b) optimized		9

An FSMD is formally defined as an ordered tuple $\langle Q, q_0, I, V, O, f : Q \times 2^S \rightarrow Q, h : Q \times 2^S \rightarrow U \rangle$, where Q is the finite set of control states, q_0 is the reset state, I is the set of input variables, V is the set of storage variables, O is the set of output variables, t is the state transition function, u is the update function of the output and the storage variables, U represents a set of storage and output assignments and S represents a set of relations over arithmetic expressions and Boolean literals.

A *path* α is a finite sequence of states where at most the first and the last states may be identical and any two consecutive states in the sequence are in f. Let $\alpha = \langle q_0 \xrightarrow{c_0} q_1 \xrightarrow{c_1} q_2 \cdots \xrightarrow{c_{k-1}} q_{l_k} \rangle$ be a path. *The condition of a path* α (C_α) is the logical expression over $I \cup V$ which is satisfied by the data state at q_0. The C_α is the weakest precondition of the path α. *The transformation of a path* α *over* V (D_α) is defined by an ordered tuple $\langle e_i \rangle$ of arithmetic expressions over the $I \cup V$ such that the value of the variable v_i is represented by e_i at the end of the path in terms of the values of the variables at the initial state of the path. There are two ways to compute the condition C_α. One is *backward* substitution, and other one is *forward* substitution. The forward substitution method relies on symbolic execution of the path.

The computation of an FSMD is defined as a finite walk starting from the start/reset state and ends again at the start state without any intermediary occurrence of the start state. It is assumed that a new computation is started when the start state is revisited and each computation terminates.

The condition and the transformation of a computation μ of an FSMD are given by C_μ and D_μ, respectively.

Let μ_1 be a computation with condition and transformation C_{μ_1} and D_{μ_1}, respectively. Also, let μ_2 be a computation with condition and transformation C_{μ_2} and D_{μ_2}, respectively. The equivalence of μ_1 and μ_2 is denoted as $\mu_1 \simeq \mu_2$ defined by $C_{\mu_1} = C_{\mu_2}$ and $D_{\mu_1} = D_{\mu_2}$. The equivalence of two paths p_1 and p_2, denoted as $p_1 \simeq p_2$, is defined in similar manner. Any computation of an FSMD F can be looked upon as concatenation of paths $[\alpha_1 \alpha_2 \alpha_3 ... \alpha_k]$ of F such that, for $1 \leq i < k$, α_i's may not all be distinct, α_i terminates in the initial state of the path α_{i+1}, the path α_1 emanates from and the path α_k terminates in the reset state q_0 of F. Hence, we have the following definition.

Definition 1 *Path cover of an FSMD* The path cover of an FSMD F is a finite set of paths P such that any computation μ of F can be presented as a concatenation of paths from P.

3.1 Equivalence of FSMDs

Let $F_0 = \langle Q_0, q_{00}, I, V_0, O, t_0, u_0 \rangle$ be the FSMD corresponding to the input of the cold scheduler and the FSMD $F_1 = \langle Q_1, q_{10}, I, V_1, O, t_1, u_1 \rangle$ be the FSMD corresponding to the output of the cold scheduler. Our objective is to show the equivalence between F_0 and F_1. In other words, we have to prove that the output values of F_0 and

F_1 are the same for all possible inputs, when the respective start states are revisited with the same values of the variables. The following definition captures the notion.

Definition 2 An FSMD F_0 is contained in an FSMD F_1, i.e., $F_0 \sqsubseteq F_1$ if $\forall \mu_0$ of F_0, $\exists \mu_1$ of F_1 such that μ_0 is equivalent to μ_1.

Definition 3 Two FSMDs F_0 and F_1 are equivalent if $F_0 \sqsubseteq F_1$ and $F_1 \sqsubseteq F_0$.

A program may contain loop, and the loop bound may not be known at compile time. To capture all possible loop iterations, we have to consider an infinite number of computations. This is not feasible to enumerate all such computations in one FSMD. The following theorem overcomes these difficulties.

Theorem 1 *For any two FSMDs F_0 and F_1, $F_0 \sqsubseteq F_1$, if there exists a finite cover $P_0 = \{p_{00}, p_{01}, \ldots, p_{0k}\}$ of F_0 for which there exists a set $P_1^0 = \{p_{10}^0, p_{11}^0, \ldots, p_{1k}^0\}$ of paths of F_1 such that $p_{0i} \simeq p_{1i}^0, 0 \leq i \leq k$.*

The equivalence checking method needs to keep correspondence between the states of two FSMDs. The notion of corresponding states can be defined as: (i) The reset states pair $\langle q_{00}, q_{10} \rangle$ is CS. (ii) Let $q_{0i} \in Q_0$ and $q_{1j} \in Q_1$ be CS. There is a path α from q_{0i} to q_{0x} in F_0. There is a path β from q_{1j} to q_{1y} in F_1 s.t. $\alpha \simeq \beta$, then $\langle q_{0x}, q_{1y} \rangle$ are CS.

3.2 Equivalence Checking Method

The algorithm 1 gives a verification method based on Theorem 1 for checking equivalence of two FSMDs.

Algorithm 1 Equivalence Checking of FSMDs

1: Insert cutpoints in F_0 and F_1.
2: Obtain the path cover P_0 and P_1 of paths of F_0 and F_1 respectively. Let P_0 be $\{p_{00}, p_{01}, \cdots, p_{0k}\}$ and let P_1 be $\{p_{10}, p_{11}, \cdots, p_{1l}\}$.
3: $\tau = \{\langle q_{00}, q_{10} \rangle\}$, where τ is the set of CS.
4: **for all** each $\langle q_{0i}, q_{1j} \rangle$ of μ **do**
5:      Show that each path $\forall p_{0x} \in P_0$ starting from the state q_{0i}, $\exists p_{1y}$ starting from q_{1j} in P_1 of F_1 such that p_{0x} is equivalent to p_{1y}.
6:      Add the end states of p_{0x} and p_{1y} as CS in τ.
7: **end for**
8: Execute the above loop by interchanging F_0 and F_1.

In this work, we select the following states as the cutpoints in an FSMD.

1. The start state of the FSMD is a cutpoint.
2. Any branching state is also cutpoint.

This choice ensures that all loops of the program are cut by at least one of the cutpoints because any loop has an exit path which creates a branching state. This idea is inherited from Floyd–Hoare's method of program verification [4].

4 FSMD Extraction from Assembly Code

The algorithm maintains a hash table of states. A state may be associated with a label as decided by the input program. During branch operation, we may have to visit a state which is already created earlier. This state can be obtained by the corresponding label. The algorithm maintains two variables $currentState$ and $nextState$. The variable $currentState$ maintains the current state of the output FSMD, and $nextState$ maintains the next state of the present transition. For non-branching operation of the form $< op > out\ in1\ in2$, we add a transition of the form $currentState - /out = in1 < op > in2\ nextState$. The present $nextState$ state becomes the $currentState$ in the next iteration in this case. For branching operation, we have two branches. We have to determine two next states one for true branch and the other for false branch.

Algorithm 2 Algorithm to convert Assembly code to FSMD

```
Create unique state for each label;
Create the start state; Put it as nextState;
while (not end of input file)
{
  read next key;
  if the key is a label
    get the state corresponding to label;
  else
    copy nextState to currentState;
  if(key is not a branch key)
    read the operations and add a transition of in FSMD;
    determine the nextState based on the next key;
  else //it is a branch key
    Determine the nextState and nextState1 from the labels
    then_part and else_part, respectively;
    Add two transitions for the branch statement;
}
```

The branch can be of the following types: branch of equal (be), branch of greater than (bg), branch of less than (bl), branch of less than equal to (ble), etc. The branch statement is followed by the label corresponding to the else part of the code and followed by no operation (nop). The then part label is followed. The two next states are determined by these two labels. If a label already exists in the state table, we get the corresponding state. Otherwise, we create a new state for the label and store in

Fig. 2 An example in
assembly code

```
Assembly Code
test:
        cmp x1 17
        bg done
        nop
        add x1 y1 x1
        add z1 1 z1
        ba test
        nop
done:
```

Fig. 3 FSMD corresponding
to assembly code in Fig. 2

```
FSMD
q0 1 - / out0 = x1 > 17 q2;
q2 2 x1 > 17 / - q1 !(x1 > 17) / -    q3;
q3 1 - / x1 = x1 + y1 q4
q4 1 - / z1 = z1 + 1 q5
q5 1 - / - q0;
q1 1 - / - q0
```

the hash table. Let us assume that $nextState$ and $nextState1$ be the two states corresponding to $then\_part$ and $else\_part$, respectively. Then, we add the following two transitions of the form corresponding to branch statement. $currentState\ 2\ in1 < be > in2 / - nextState\ !(in1 < be > in2) / - nextState1$. The $nextState$ becomes the next $currentState$ to proceed with. The abstract algorithm for converting an assembly code to its equivalent FSMD is given as algorithm 2.

An assembly code in text format is given in Fig. 2. The corresponding FSMD in text format generated by our algorithm is given in Fig. 3. For visualization of the FSMD, we have written another parser to convert the FSMD to its equivalent dotty format. The dotty file can be viewed as a graph in Graphviz software. Also, dotty can be converted automatically to pdf for viewing in any pf viewer like acrobat reader. This graphical representation helps the user to visualize a behavior and debug any non-equivalence easily.

5 Verification of Cold Scheduling

The FSMD extracted from the assembly codes of Table 1 is shown in Fig. 4. There is only path in both the FSMDs. The verification tasks here are to show that the condition and the transformation of the respective paths are equivalent. Let us assume that the path in the initial FSMD be $\alpha = \langle q00 \rightarrow q08 \rangle$ and the path in the output FSMD be $\beta = \langle q10 \rightarrow q18 \rangle$. The condition of both the paths is TRUE, i.e., $C_\alpha = C_\beta = TRUE$. The transformation of the path α is $D_\alpha = \langle b * b - 4 * a * c, a * c, 4 * a * c, c, 4 \rangle$ where the order of the registers is $R0$, $R1$, $R2$, $R3$. Similarly, the transformation of the path β is $D_\beta = \langle b * b - 4 * a * c, a * c, 4 * a * c, c, 4 \rangle$, where

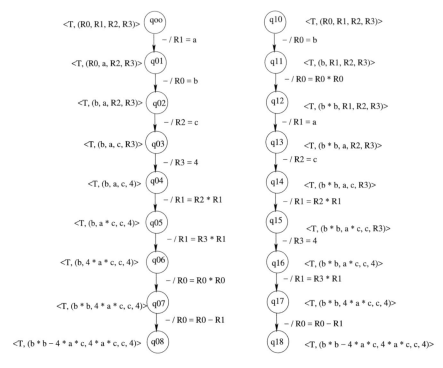

Fig. 4 FSMD corresponds to the assembly code of Table 1 with transformation

the order of the registers is $R0$, $R1$, $R2$, $R3$. The computation of transformation is shown stepwise in Fig. 4. This computation is based on symbolic simulation. Clearly, transformation of both the paths is equivalent, i.e., $D_\alpha = D_\beta$. Hence, both the FSMDs in Fig. 4 are equivalent.

We took a simple example to explain the logic of our method clearly. However, in the practical case, the behavior may contain thousand lines of code, conditional statement and loops. So, the corresponding FSMD will contain many paths. In this case, manually proving the equivalence between the FSMDs will be impossible. In that case, our path-based equivalence checking method will play a crucial role in proving correctness of cold scheduling.

6   Experimental Results

We have implemented our verification framework in C. To create test case, we have taken some examples of SPARC assembly language from [6]. Since *SPARC* uses RISC, it is very clean, simple and still being powerful. The test cases are run on an Intel Core I5 7200U Processor machine with 4GB RAM. The results are shown in

Table 2 Experimental results

Benchmarks	#operation	#variable	#If-else	#loop	State in FSMD	Runtime (in ms)
$Test\_1$	10	2	1	1	9	1.4
$Test\_2$	11	2	1	1	10	1.4
$Test\_3$	5	1	1	0	5	0.8
$Test\_4$	7	3	1	0	7	1.0
$Test\_5$	16	3	1	1	15	1.8

Table 2. As shown in table, we have considered test cases with if-else and also with loops. In our experimental setup, we have generated optimized version of SPARC assembly language through cold scheduling algorithm given in [7]. After that, we have generated FSMD from both the source and the optimized version of codes through our assembly to FSMD parser. These two FSMDs are the input to our equivalence checking method. As shown in table, our method takes very little time to establish the equivalence. This shows the usefulness of our method.

7 Conclusions

A framework to ensure correctness of cold scheduling is presented here. Our method first extracts FSMD from assembly-level code and then applies an equivalence checking method to prove the equivalence. We have also presented a visualization framework of FSMD to ease the debug process of the users. The method is implemented and tested on some examples available online. This is the first work which targets verification of cold scheduling. As a future work, we want to verify other energy-efficient transformation like register relabeling.

References

1. Graphviz - graph visualization software. https://graphviz.gitlab.io/
2. D. Gajski, L. Ramachandran, Introduction to high-level synthesis. IEEE Trans. Des. Test Comput. pp. 44–54 (1994)
3. D. Grune, K.V. Reeuwijk, H.E. Bal, C.J. Jacobs, K. Langendoen, *Modern Compiler Design* (Springer, Berlin, 2012)
4. C.A.R. Hoare, An axiomatic basis for computer programming. Commun. ACM **12**, 576–580 (1969). https://doi.org/10.1145/363235.363259
5. C. Karfa, D. Sarkar, C. Mandal, P. Kumar, An equivalence-checking method for scheduling verification in high-level synthesis. IEEE Trans. Comput.-Aided Des. Integr. Circuits Syst. **27**(3), 556–569 (2008)
6. S. Microsystems, Ksparc assembly language - clemson university. https://people.cs.clemson.edu/~mark/sparc/control_structures.txt

7. C.L. Su, A.M. Despain, Cold scheduling. Tech. rep. (1993)
8. J.B. Tristan, X. Leroy, Verified validation of lazy code motion. SIGPLAN Not. **44**(6), 316–326 (2009)
9. T.H. Weng, C.H. Lin, J.J.J. Shann, C.P. Chung, Power reduction by register relabeling for crosstalk-toggling free instruction bus coding, in *2010 International Computer Symposium (ICS2010)*, pp. 676–681 (2010)

A Systematic Review on Program Debugging Techniques

Debolina Ghosh and Jagannath Singh

Abstract In software engineering, debugging is a most tedious job. Finding and correcting the bug takes much more time and effort than coding. Many researchers have worked for making the debugging process easier. Many existing debugging techniques are available. Here, in this paper, we review various new emerging trends of software debugging techniques which is mostly used by the developers or testers for a particular application.

Keywords Debugging · Testing · Java · Breakpoint · Omniscient debugging

1 Introduction

Nowadays, technology is developing so fast and everything is dependent on software. So, the size and complexity of the software are increasing exponentially. It is obvious that debugging or testing such a large and complex software takes more time and effort when performed manually. Many researchers are working on debugging technique to make the debugging process easier. For fast and efficient debugging, it is necessary to reduce the size of the program by eliminating the number of statements that are not causing the error. Debugger can inspect the state of the executing program and understand its impact on program's behavior.

In this paper, we review some of the existing debugging techniques and their uses. The paper is structured as follows: Sect. 2 describes the background details, and Sect. 3 contains the different types of existing debugging methods. Section 4 concludes the paper.

D. Ghosh · J. Singh (✉)
School of Computer Engineering, KIIT Deemed to be University, Bhubaneswar 751024, India
e-mail: jagannath.singhfcs@kiit.ac.in

D. Ghosh
e-mail: debolina442@gmail.com

© Springer Nature Singapore Pte Ltd. 2020
A. Elçi et al. (eds.), *Smart Computing Paradigms: New Progresses and Challenges*,
Advances in Intelligent Systems and Computing 767,
https://doi.org/10.1007/978-981-13-9680-9_16

2   Background

Korel et al. [1] have developed a dynamic slicing tool which helps the programmer in debugging process by providing some extra functionalities like executable slices, influencing variables, contributing nodes, and partial dynamic slicing. Debugger can fix a breakpoint, and when the breakpoint is reached during the program execution, the program is suspended. Then, the programmer can analyze the program state. This follows the traditional debugging process. Krebs et al. [2] have presented a software debugging tool which is used to help a programmer in locating the errors in coding at the client side using scripts or dynamic HTML.

Wong et al. [3] presented an augmentation method and a refined method which are used to overcome the debugging problem. According to this paper, these two methods can help in effective fault localization by better prioritized code based on its likelihood of containing program bug . Zhang et al. [4] have developed the systematic use of dynamic slicing tool (Jslice) to locate the Java program's fault based on dynamic slicing analysis. Singh et al. [5] have developed a tool D-AspectJ for aspect-oriented programs to compute the dynamic slices. Dynamic program slicing is used to analyze the effect of one statement on other statements.

Treffer et al. [6] have developed a debugging tool Slice Navigator, for Java programs that combine dynamic slicing with back-in-time debugging to support the debugging process. Back-in-time or omniscient debugging follows the overhead of frequently restarting the debug session. The authors represent a new approach that combines omniscient debugging and dynamic slicing. When developers omnisciently debug a dynamic slice, at any point they can add or adjust the slicing criteria and changes are applied instantly, without interrupting the debug session.

Fritzson et al. [7] developed a generalized algorithmic debugging and testing (GADT) method tool which takes the help of program slicing and extended version of category partition testing technique for its bug localization. Algorithmic debugging is an interactive process between the debugging tool and programmer where the debugging tool acquires the knowledge about the expected behavior of the executed program and uses this knowledge for bug localization. System collects this knowledge through yes or no questions to the programmer.

Silva et al. [8] developed a declarative debugger for Java (DDJ) which is a new implementation of algorithmic debugging for making the debugging process scalable. The main disadvantage of algorithmic debugger is its scalability problem. DDJ implements an algorithm that allows strategies to work with uncompleted execution trace (ET) nodes to solve the time scalability problem. Gonzlez et al. [9] developed a new hybrid debugging architecture by combining trace-based debugging, omniscient debugging and algorithmic debugging. It overcomes the drawback of other debugging techniques.

Mariga et al. [10] presented a review about the automated tools that can locate and correct the faulty statements in a program. The tool can also significantly reduce the cost of software development and improve the overall quality of the software by reducing the human effort in the process of debugging. The author discussed the fault localization, program slicing, and delta debugging techniques.

Campos et al. [11] have presented a tool-set for automatic testing and fault localization. GZoltar is an Eclipse plug-in that provides a technique for test suite reduction and prioritization. The GZoltar implements a technique called spectrum-based fault localization (SFL). SFL is based on monitoring a program to keep track of executed parts.

3 Debugging Techniques

Debugging is a process of finding out the incorrect execution and locating the root cause of the incorrectness. Different existing debugging techniques are applied for bug localization. We present in Fig. 1 a hierarchical model of all types of debugging process available in our literature survey.

3.1 Trace-Based Debugging

Trace-based debugging is a traditional debugging process that works on breakpoint concept. Breakpoint [12] works as pausing or stopping point to examine the state of a program execution. After rectifying the current bug, set the next breakpoint and repeat the same process till all the bugs are rectified.

3.1.1 Trace Debugging (TD)

Trace debugging [1, 13] is a step-by-step process. Debugger can set a breakpoint, and when a breakpoint is reached during program execution, the program is suspended. Then, the programmer can examine the program state line by line. Debugger takes only the control over the interpreter. So, the scalability of TD is same as interpreter. The abstraction level of this debugging process is very low, and it is a manual process. During bug localization, if bug is found before the breakpoint, then the debugging process needs to restart again.

3.1.2 Omniscient Debugging (OD)

Omniscient debugging [14] or back-in-time debugging [6] is a debugging process by using breakpoints. From functional point of view, it is same as trace debugging. The concept of omniscient debugging is that a debugger can not only rewind time, but also has instant access to every point in past and future. It can trace the computation from both backward and forward ways. The programmer can go back to find out the

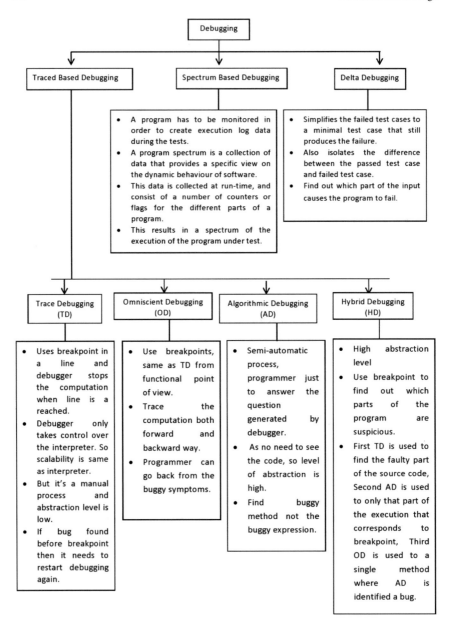

Fig. 1 Classifications of debugging

buggy symptoms as it traces the execution tree with time. But it has a low level of abstraction, and it is a manual process. In this type of debugging, execution traces are huge and storing them is a challenging task, so the scalability of the omniscient debugging is low.

3.1.3 Algorithmic Debugging (AD)

Algorithmic debugging [7, 8] is a semiautomatic debugging system that produces a dialog between the debugger and the developer to find bugs. So, there is no need to see the code. For that reason, the level of abstraction is high for algorithmic debugging. In this type of debugging, there are two phases: In the first phase, it builds the execution tree, and in the second phase, the execution tree is explored. Execution tree is generated according to the strategies like top-down, single stepping, heaviest first, and divide and query. The main drawbacks of algorithmic debugging are that the execution tree needs much more main memory and construction of ET is also high. This type of debugging finds the buggy method, not the buggy expression.

3.1.4 Hybrid Debugging (HD)

Hybrid debugging technique [9] is a combination of trace debugging, omniscient debugging, and algorithmic debugging. First, it uses the traditional trace debugging method using breakpoint to find out the suspicious part of the program. Then, it uses algorithmic debugging to only that part of the execution that corresponds to break-point and then omniscient debugging is used to a single method where algorithmic debugging identified a bug.

3.2 Spectrum-Based Debugging

Spectrum-based debugging [11] or spectrum-based fault localization (SFL) [12] is the process of monitoring the statements involved in a particular execution trace. Program spectrum is the state which identifies the active part of the program during its run. Fault localization is the process to identify the part of a program which may cause the error. Program spectrum is a collection of data at runtime that gives a specific view about the dynamic behavior of the software. It consists of counters or flags for the different parts of the program. There are different types of program spectra exist such as block hits and function hits.

3.3 Delta Debugging

Delta debugging [15, 16] is a process of automated test case minimization. It takes test cases that may cause the error and prepare the error report. Based on the error report, minimal test cases are selected that are significant for producing failure. The minimal test cases will generate the same error, and hence, it will be helpful for bug localization. Delta debugging technique automates this process of repeated trails for minimized input. For example, if we can provide a test case that may cause the bug, then we apply delta debugging algorithm. It will try to minimize the irrelevant source codes and functions to reproduce the bug, until a minimal program is found.

$$y \neq x[15] \tag{1}$$

4 Conclusion

In this paper, we study some existing debugging techniques which are mostly used by the developers or testers. Trace-based debugging using breakpoint is the traditional method and manual process. Though algorithmic debugging and hybrid debugging are semiautomatic processes, these processes use the breakpoint to examine the program state. Spectrum-based debugging is used to analyze the runtime behavior of the program or the program spectra like basic blocks or functions. Delta debugging is the automatic test case minimization process which takes the input test cases and provides the only test cases which are significant for producing the bug.

References

1. B. Korel, J. Rilling, Application of dynamic slicing in program debugging, in *Proceedings of the 3rd International Workshop on Automatic Debugging (AADEBUG-97)*, No. 1 (Linkping University Electronic Press, 1997)
2. W.H. Krebs, M.W. Lumsden, Software debugging tool for displaying dynamically written software code. U.S. Patent No. 6668369 (2003)
3. W.E. Wong, Y. Qi, Effective program debugging based on execution slices and inter-block data dependency. J. Syst. Softw., **79**(7), 891–903 (2006)
4. P. Zhang, Fault localization based on dynamic slicing via JSlice for Java programs, in *5th IEEE International Conference on Software Engineering and Service Science (ICSESS)*. IEEE (2014)
5. J. Singh, P.M. Khilar, D.P. Mohapatra, Dynamic slicing of distributed aspect-oriented programs: a context-sensitive approach. Comput. Stand. Interfaces **52**, 71–84 (2017)
6. A. Treffer, M. Uflacker, The slice navigator: focused debugging with interactive dynamic slicing, in *IEEE International Symposium on Software Reliability Engineering Workshops (ISSREW)*. IEEE (2016)
7. P. Fritzson, Generalized algorithmic debugging and testing. ACM Lett. Program. Lang. Syst. (LOPLAS) **1**(4), 303–322 (1992)

8. I. David, J. Silva, An algorithmic debugger for Java, in *IEEE International Conference on Software Maintenance (ICSM)* (2010), pp. 1-6
9. J. Gonzlez, D. Insa, J. Silva, A new hybrid debugging architecture for eclipse, in *LOPSTR* (2013), pp. 183–201
10. G. Mariga, K. Mwiti, *Automatic Debugging Approaches: A Literature Review* (Murang'a University College, Muranga, Kenya, 2017)
11. J. Campos, A. Riboira, A. Perez, R. Abreu, Gzoltar: an eclipse plug-in for testing and debugging, in *Proceedings of the 27th IEEE/ACM International Conference on Automated Software Engineering* (2012), pp. 378–381
12. T. Janssen, R. Abreu, A.J.C. Van Gemund, Zoltar: a spectrum-based fault localization tool, in *Proceedings of the ESEC/FSE Workshop on Software Integration and Evolution* (ACM, 2009), pp. 23–30
13. D. Hao, L. Lingming Zhang, J. Sun Zhang, H. Mei, VIDA: visual interactive debugging, in *31st International Conference on Software Engineering* (ICSE). IEEE (2009), pp. 583–586
14. G. Pothier, E. Tanter, J. Piquer, Scalable omniscient debugging. ACM SIGPLAN Not. **42**(10), 535–552 (2007)
15. C. Artho, Iterative delta debugging. Int. J. Softw. Tools Technol. Transf. **13**(3), 223–246 (2011)
16. M. Burger, K. Lehmann, A. Zeller, Automated debugging in eclipse, in *Companion to the 20th Annual ACM SIGPLAN Conference on Object-Oriented Programming, Systems, Languages, and Applications* (2005), pp. 184–185

Big Data and Recommendation Systems

Protein Sequence Classification Involving Data Mining Technique: A Review

Suprativ Saha and Tanmay Bhattacharya

Abstract In the field of bio-informatics, size of the bio-database is increasing at an exponential rate. In this scenario, traditional data analysis procedure fails to classify it. Currently, a lot of classification techniques involving data mining are used to classify biological data, like protein sequence. In this paper, most popular classification techniques, like neural network-based classifier, fuzzy ARTMAP-based classifier, and rough set classifier are reviewed with the proper limitation. The accuracy level and computational time are also been analyzed in this review. At the end, an idea is proposed which can increase the accuracy level with low computational overhead.

Keywords Data mining · Neural network · Fuzzy ARTMAP · Rough set · String kernel · Protein-hashing · SVM/GA

1 Introduction

Nowadays, large amount of biological data are produced from the research in proteomics, genomics, and other biological domains. Instead of traditional data analysis system, data mining technique is used to interpret the voluminous biological data. Protein structure prediction, gene classification, analysis of mutations in cancer, and gene expressions are the popular areas of research involving data mining in the field of bio-informatics. In this scenario, to classify protein sequence, different classification approaches were invented from time to time. This paper presents a brief review of the different classification methods in Sect. 2. Section 3 presents a comparative study of limitation between existing classifier, followed by a conclusion in Sect. 4.

S. Saha (✉)
Department of Computer Science and Engineering, Brainware University,
Barasat, Kolkata, India
e-mail: reach2suprativ@yahoo.co.in

T. Bhattacharya
Department of Information Technology, Techno India, Salt Lake, Kolkata, India
e-mail: dr.tb1029@gmail.com

© Springer Nature Singapore Pte Ltd. 2020
A. Elçi et al. (eds.), *Smart Computing Paradigms: New Progresses and Challenges*,
Advances in Intelligent Systems and Computing 767,
https://doi.org/10.1007/978-981-13-9680-9_17

203

2   Related Work

Various researchers proposed different classification approaches to classify the unknown protein sequence from time to time, which were mentioned in this section with their proper accuracy level of classification.

2.1   Neural Network Model Based Classifier

A sequence classified based on neural network model was proposed by Jason T. L. Wang et al. [1]. In this approach, features were extracted from the protein sequence using '2-gram encoding method' and '6-letter exchange group methods', which find the global similarity of the protein sequence. Some user-defined variables like 'Len', 'Mut', and 'occur' were also used to find the local similarity of the protein sequence. Minimum description length (MDL) principle was also used to calculate the significance of motif. Cathy Wu et al. [2] enlarged 2-gram encoding method to n-gram encoding method to increase the accuracy level of classification. Zarita Zainuddin et al. [3] had proposed the radial-based approach to reduce the computational overhead of n-gram encoding method. PV Nageswara Rao et al. [4] proposed self-organized map (SOM)-based probabilistic neural network model. The 90% accuracy was evaluated by this model.

2.2   Fuzzy ARTMAP Model Based Classifier

Shakir Mohamed et al. [5] proposed fuzzy ARTMAP classified, which has some advantage over neural network based model with respect to data analysis. At first, noisy data is removed from the input, then the physic–chemical properties of the protein sequence, i.e., molecular weight (W), isoelectric point (pI), hydropathy composition (C), hydropathy distribution (D), and the hydropathy transmission (T) are calculated as feature values. This model was claimed 93% accuracy level of classification. E. G. Mansoori et al. [6] applied the result of 6-letter exchange group method on the distributed pattern matrix to increase the accuracy level. After that, it had been tried to decrease the computational time without changing the classification accuracy [7]. Finally, this model was compared with non-fuzzy technique.

2.3   Rough Set Classifier

Instead of handling large number of unnecessary features, Shuzlina Abdul Rahman et al. [8] had proposed a procedure to select the necessary feature using rough set approach. Rough set theory is a machine learning method, for taking decision, which is introduced by Pawlak [9]. A faster, approx. 97% accurate classification model involving rough set theory was proposed by Ramadevi Yellasiri et al. [10]. Sequence arithmetic, rough set theory, and concept lattice were used to develop this model. Among the knowledge-based analysis and analysis of data, rough set classifier model only provides knowledge-based information. To increase the accuracy level, structural analysis of the protein sequence is also applicable.

2.4 Model Based on the Combination of Fuzzy ARTMAP and Neural Network Model

Feature grouping approach based on the characteristics and implementation policy of features was proposed by Saha S. et al. [11]. This approach was applied to a model which is a combination of neural network system, fuzzy ARTMAP model, and rough set classifier. The KMP string matching algorithm was used instead of window sliding technique in the phase of neural network model to improve the computational time. This model was also reached to 91% accuracy level of classification.

2.5 String Kernel Method Based Classifier

Detection of both functional and structural similarities between protein sequences is the key idea of protein sequence classification. Different types of approaches, i.e., 'local pair-wise methods', 'profile hidden Markov model based methods' and 'recursive method' were used to identify the structural similarity of protein. Spalding J. D. et al. [12] had proposed the string kernel method-based classifier to develop a strategy for efficient estimation of suitable kernel parameter values. Here, the Kullback–Leibeir distance was calculated between the observed k-mar frequencies and the theoretical k-mar frequencies of protein data which was generated from a null-model of k-mar generation. This classifier was provided 87.5% accuracy.

2.6 String Weighting Scheme-Based Classifier

N. M. Zaki et al. [13] proposed a technique to extract the feature value using hidden Markov model, which was applied to the classifier that can train the data in high-dimensional space. This approach was also used support vector machines (SVM) based classified to learn the boundaries between structural protein classes. With the combination of feature extraction phase and classification phase were formed string weighted based classifier which was claimed 99.03% accuracy. Finally, this classifier was compared with other five existing classifiers.

2.7   Classifier Involving Fourier Transform

A. F. Ali et al. [14] proposed a classification approach based on the functional properties of the protein sequence. In this technique, a set of spectral domain features based on fast fourier transform were applied on SCOP database. This classifier had utilized multilayer back propagation (MLBP) neural system for protein classification. The classification accuracy was 91% when it was applied on full four levels of SCOP database. The accuracy level was enhanced up to 96% when it was applied on restricted level of SCOP.

2.8   Tree-Based Classifier

Robert Busa-Fekete et al. [15] proposed the phylogenetic analysis approach in the classification of protein sequence data. This technique was the combination of two algorithms, e.g., (a) TreeInsert algorithm was responsible to determine the necessary minimum cost which was inserted in phylogenetic tree, (b) TreNN [16] algorithm was assigned the label to the query based on an analysis of the query's neighborhood approach. At the end, this model was compared with different classification models like BLAST, Smith-Waterman, and local alignment kernel and was achieved 93% accuracy of classification.

2.9   Hidden Markov Based Classifier

Pranay Desai [17] had proposed a classification technique based on hidden Markov models (HMMs) to classify protein sequences. HMMs were used three phases, e.g., training, decoding, and evaluation to classify the functional properties of input data. The sequence data was converted into the training cluster in "training phase" as well as "decoding" and "evaluation phase" was provided likelihood analysis approach. 94% accurate classification had been given by this model.

2.10   Classifier Using Structural Analysis of Protein

Muhammad Mahbubur Rahman et al. [18] proposed a classification approach based on hierarchy tree structure. This technique was provided the solution of feature grouping problem. The PSIMAP technique [19] was very well-known approach to feature grouping, but it also failed to classify scale-free protein. This method involving six major attributes of proteins like (a) structure comparison, (b) sequence comparison, (c) connectivity, (d) cluster index, (e) interactivity, and (f) taxonomic was solved the previous drawback with 98% accuracy level. Binary search technique was also applied here.

2.11 Classifier Using Feature Hashing

Cornelia Caragea et al. [20] were proposed a method to reduce the complexity of learning algorithm where input belongs in high-dimensional space. The most popular feature extraction procedure, e.g., K-gram encoding method, produces high-dimensional data for the large value of k. This high-dimensional space was reduced in low by feature hashing mechanism. Here, more than one features were mapped to the same hash key and aggregating their count. This procedure was also compared with "bag of k-gram" approach. At the end, this feature hashing classifier was reached to 82.83% accuracy.

2.12 Classifier Using GA/SVM System

Xing-Ming Zhao et al. [21] proposed a classifier with a combination of genetic algorithm (GA) and support vector machine (SVM) framework. Features were extracted by genetic algorithm and that had been trained by SVM. This model was applied to PIR protein database [6] and classified the protein sequence with 99.24% accuracy. At the end, this approach was compared with BLAST and HMMer classifier.

3 Comparative Study Between Limitation of the Existing Classifiers

Table 1 shows the comparative analysis with the accuary and limitation between the existing classifier to classify unknown protein sequences into its families.

4 Conclusion

The knowledge extraction from the unprocessed data is very important in biological domain. In this case to find the accurate classifier, to classify unknown protein sequence into its proper family, is an important area of research. In this paper, working principle of different classification approaches is reviewed with the accuracy level of classification. From this review, it can be concluded that no classification model is maintained highest level of accuracy with acceptable computational time

Table 1 Comparative study between existing classifiers with respect to limitation and accuracy

Name of classifier	Accuracy (%)	Limitation
Neural network-based classifier [1–4]	90	Linearity problem degrades the accuracy level
Fuzzy ARTMAP classifier [5–7]	92	Requires more storage and time
Rough set classifier [8–10]	97.7	Only knowledge-based information is extracted
Model in a combination of fuzzy ARTMAP and neural network model [11]	91	Structural analysis is missing, so accuracy level is not up to the mark
String kernel method-based classifier [12]	87.5	
String weighting scheme-based classifier [13]	99.03	
Classifier involving Fourier transform [14]	96	
Tree-based classifier [15, 16]	93	
Hidden Markov-based classifier [17]	94	
Classifier using feature hashing [20]	82.83	
Classifier using structural analysis of protein [18]	98	Individual structure of a protein sequence is not analyzed
Classifier using GA/SVM system [21]	99.24	Global similarity is not implemented

at a same time. After analyzing techniques mentioned above, it can be finalized that local structural analysis means position analysis of the protein sequence helps to increase the accuracy level without disturbing the computational time.

References

1. T.L. Jason et al., *Application of Neural Networks to Biological Data Mining: A case study in Protein Sequence Classification* (KDD, Boston, 2000), pp. 305–309
2. C. Wu, M. Berry, S. Shivakumar, J. Mclarty, *Neural Networks for Full-Scale Protein Sequence Classification: Sequence Encoding with Singular Value Decomposition* (Kluwer Academic Publishers, Boston, Machine Learning, 1995), pp. 177–193
3. Z. Zainuddin, M. Kumar, Radial basic function neural networks in protein sequence classification. MJMS **2**(2), 195–204 (2008)
4. P.V. Nageswara Rao, T. Uma Devi, D. Kaladhar, G. Sridhar, A.A. Rao (2009) A probabilistic neural network approach for protein superfamily classification. J. Theor. Appl. Inf. Technol
5. S. Mohamed, D. Rubin, T. Marwala, in *Multi-class Protein Sequence Classification Using Fuzzy ARTMAP*. IEEE Conference. (2006) pp. 1676–1680
6. E.G. Mansoori et al., Generating fuzzy rules for protein classification. Iran. J. Fuzzy Syst. **5**(2), 21–33 (2008)
7. E.G. Mansoori, M.J. Zolghadri, S.D. Katebi, Protein superfamily classification using fuzzy rule-based classifier. IEEE Trans. Nanobiosci. **8**(1), 92–99 (2009)

8. S.A. Rahman, A.A. Bakar, Z.A.M. Hussein, in *Feature Selection and Classification of Protein Subfamilies Using Rough Sets*. International Conference on Electrical Engineering and Informatics. (Selangor, Malaysia, 2009)
9. Z. Pawlak (2002) Rough set theory and its applications, J. Telecommun. Inf. Technol
10. R. Yellasiri, C.R. Rao, Rough set protein classifier. J. Theor. Appl. Inform. Technol (2009)
11. S. Saha, R. Chaki (2012) Application of data mining in protein sequence classification. IJDMS. 4(5)
12. J.D. Spalding, D.C. Hoyle, *Accuracy of String Kernels for Protein Sequence Classification, ICAPR 2005*, vol. 3686. (Springer (LNCS), 2005)
13. N.M. Zaki, S. Deri, R.M. Illias, Protein sequences classification based on string weighting scheme. Int. J. Comput. Internet Manage. **13**(1), 50–60 (2005)
14. A.F. Ali, D.M. Shawky, A novel approach for protein classification using fourier transform. IJEAS **6**(4), 2010 (2010)
15. R. Busa-Fekete, A. Kocsor, S. Pongor (2010) Tree-based algorithms for protein classification. Int. J. Comput. Sci. Eng. (IJCSE)
16. K. Boujenfa, N. Essoussi, M. Limam, Tree-kNN: A tree-based algorithm for protein sequence classification. IJCSE **3**, 961–968 (2011)
17. P. Desai, Sequence Classification Using Hidden Markov Model (2005)
18. M.M. Rahman, A.U. Alam, A. Al-Mamun, T.E. Mursalin, A more appropriate protein classification using data mining. JATIT, 33–43 (2010)
19. D. Bolser et al., Visualization and graph-theoretic analysis of a large-scale protein structural interactome. BMC Bioinformatics **4**, 1–11 (2003)
20. C. Caragea, A. Silvescu, P. Mitra, Protein sequence classification using feature hashing. Proteome Sci. **10**(Suppl 1), S14 (2012)
21. X.M. Zhao et al., A Novel Hybrid GA/SVM System for Protein Sequences Classification, IDEAL 2004, vol. 3177. (Springer(LNCS), 2004), pp. 11–16

Cache Memory Architectures for Handling Big Data Applications: A Survey

Purnendu Das

Abstract Cache memory plays an important role in the efficient execution of to-day's big data-based applications. The high-performance computer has multicore processors to support parallel execution of different applications and threads. These multicore processors are placed in a single chip called chip multiprocessor (CMP). Each core has its own private cache memories, and all the core share a common last-level cache (LLC). The performance of LLC plays a major role in handling big data-based applications. In this paper, we have done a survey on the innovative techniques proposed for efficiently handling big data-based applications in the LLC of CMPs.

Keywords Cache memory · Near-data processing · Big data · Multicore architectures

1 Introduction

Cache memory is one of the essential components for all the modern processors. In today's processors, multiple cores are placed in a single chip (or die) which is called the chip multiprocessor (CMP) [1]. In CMP, each core has its own cache memories, and all the cores also share a common large size last-level cache (LLC). Figure 1 shows an example of such CMP having four cores and two levels of cache memories. Each core has its own private L1 cache, and all the cores share an L2 cache as LLC. As the data demand of the applications increases, the cache architectures are also needed to be redesigned for supporting such big data-based applications [2, 3]. The details about basic cache organization is discussed in Sect. 2.

A simple technique to handle big data is to increase the cache size. But increasing cache is not very simple. A large size cache may improve the number of hits in the cache, but searching time of a block within the cache will increase, hence increases the

P. Das (✉)
Department of Computer Science, Assam University Silchar, Silchar, Assam, India
e-mail: purnen1982@gmail.com

© Springer Nature Singapore Pte Ltd. 2020
A. Elçi et al. (eds.), *Smart Computing Paradigms: New Progresses and Challenges*,
Advances in Intelligent Systems and Computing 767,
https://doi.org/10.1007/978-981-13-9680-9_18

Fig. 1 An example of chip multiprocessor (CMP)

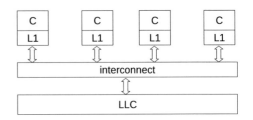

cache access latency. Multiple levels of cache are the solution for this, but such kinds of designs are very old and all the commercial processors today have multiple levels of cache memories [4, 5]. Hence, more advanced technology is required to handle such issues. In this paper, we have done a survey of the recent research activities performed to improve the cache performance. Our main target for this survey is the LLC. Research has done in many directions to improve the performance of LLC. The two most important directions are *reducing cache access latency* and *reducing miss rate*.

Reduction of miss rate means more number of hits into the cache which improves the performance of the system. A data-intensive application with high temporal locality will be beneficial with such reduction of cache misses. There are many techniques proposed to reduce the miss rates [6–9]. Reducing access latency of large-sized LLC is also a major area of research [10–13]. The cache access latency increases if larger cache is used to reduce the miss rate. But smaller cache size increases the miss rate. Hence, designing a better cache architecture is challenging, and it is a major area of research [1, 3, 14, 15]. The main goal of proposing better LLC architecture is that the performance of the system should improve. The performance normally measures in terms of cycles per instruction (CPI) or instructions per cycle (IPC) [1].

The organization of this paper is as follows. The next section discusses about the basic cache organization. Section 3 discusses the techniques proposed for reducing the LLC access latencies. Section 4 discusses about the reduction of miss rate in LLC. A special technique called near-data processing (NDP) for handling big data-based applications efficiently is discussed in Sect. 5. Finally, Sect. 6 concludes the paper.

2 Cache Organization

Most of the today's cache memories are set-associative cache where the cache has multiple sets, and each set has a fixed number of ways. The set-associative cache can be compared with a two-dimensional array where the sets can be called as rows and the ways as columns. Each entry of the array stores a cache block. A block can always map to a fixed set based on the *set index* from its address bits. Some bits from the address of each block are reserved as *set index*. The block can be placed in anyways (column) of the corresponding set. If the set is full, the replacement policy selects a victim block to replace it with the new block. To search a block within the cache, all the ways in the corresponding set are to be searched simultaneously. The detail about the cache organization is mentioned in [15].

3 Reducing Access Latency

Cache memories were initially accessed with uniform access latency (UCA) [1]. Such uniform cache works fine when the cache is small; for todays, large size caches (especially LLC), such as UCA-based caches, create latency overhead.

3.1 NUCA

To reduce the latency overhead of UCA, in [11] the cache is divided into multiple banks. Each bank acts as an independent set-associative cache, and all the banks are connected with some on-chip networks. The on-chip network or network on chip (NoC) is used as the alternative of the bus to communicate among the different on-chip modules. Figure 2a shows an example of such NUCA architecture. The CMP shown in the figure has four cores, and the LLC is divided into 16 banks. Note that, the other upper levels of cache memories except LLC are not shown in the figure. The main advantage of NUCA is each bank can be accessed independently, and the nearby banks can be accessed much faster than the farther banks. For example, in the figure, if the fourth core wants to access a block from *bank-15*, then it can access it much faster than a block from *bank-0*. Some on-chip communication is required to access *bank-0*. Most of the today's LLC use NUCA-based architectures [4, 14, 16–18]. Proposing NUCA allows to use large size LLC without any major latency overhead, which is essential for running big data-based applications.

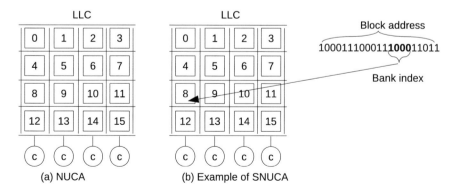

Fig. 2 NUCA. **a** An example of NUCA. **b** An example of SNUCA

There are two types of NUCA architectures available: Static NUCA (SNUCA) and Dynamic NUCA (DNUCA). The physical organization of both SNUCA and DNUCA is the same; they differ in the data (block) management policy. In SNUCA, a block can only be mapped to a fixed bank based on its *bank index* bits. Same as *set index* some bits from the block address are used for the *bank index* and based on these bits a block always maps to a fixed bank. For example, in Fig. 2b, the block maps to bank 8 based on the *bank index* **1000**. SNUCA is simple and easy to implement. The block searching mechanism is very easy as a block always maps to a fixed bank. Due to its simple data management policy, SNUCA is used by many technologies proposed [7–9, 19, 20].

The major disadvantage of SNUCA is that it cannot migrate block from one bank to another. For example, in Fig. 2a, the core needs longer time to access data from *bank-0* than *bank-15*. Suppose there is a block **A** resides in *bank-0* which is requesting by the core multiple times then in SNUCA every time the block has to travel through the on-chip network to reach the core. It would be better if the block can be migrated to a closer bank. Such migration is not allowed in SNUCA. Also due to fixed mapping, the banks may also show some other non-uniform behaviors [7, 21].

The DNUCA allows to migrate data from one bank to another. DNUCA groups the banks into multiple *bankset*. For example, each column in Fig. 2a can be considered as a bankset. A block can be placed in any of the bank within the bankset. Hence, a block can be moved from one bank to another (within the bankset) according to the requirement. DNUCA-based architecture is also beneficial for removing other non-uniform bank behaviors [22, 23]. The main operations of DNUCA are

- **Searching**: To search a block, the entire bankset has to be searched. Searching a block in DNUCA is time-consuming, and it must be optimized [1, 24].
- **Migration**: A heavily used block from one bank can be migrated to a closer bank [10].

Though DNUCA has been proved as the better architecture than SNUCA, proper implementation of DNUCA has some challenges. Researchers are still working on improving the searching time, reducing migration overhead, etc. The next section discusses about the issues with DNUCA and some of the innovative techniques proposed to handle such issues.

3.2   Issues with DNUCA

The main issue of DNUCA is the block searching time. Since all the banks within the bankset have to be searched, optimized searching policy must be used for better performance. There are three basic searching policies used in DNUCA:

- **Incremental**: Search the banks sequentially one after another. This technique less complicated and also less number of network communication is required. But it may increase the search latency; in the worst case, all the four banks have to be searched.

- **Multi-cast**: In multi-cast search, all the banks are searched simultaneously. This technique is fast, but due to multi-cast, it may increase the network communication overhead which may lead to congestion and high power consumption [1].
- **Mixed**: In mixed search, some of the banks are searched simultaneously, and remaining banks are then searched incrementally.

There are lot of other searching techniques have been proposed [3, 24–26]. A smart searching technique is proposed in [3]. Instead of searching all the banks, the proposed technique maintains a record (partial tag record) based on which less number of banks is required to search. For example, if the bankset has eight banks, then the smart searching technique may guide to search only four banks. Such smart searching technique requires an additional storage to store the partial tag information of each block in the cache [12].

As mentioned before, each bank in NUCA is a set-associative cache. In set-associative cache, each block must be placed to its fixed set based on its *set index*. Such property of set-associative cache limits the power of DNUCA migration [12]. In DNUCA, the heavily used blocks can be migrated to a closer bank. If many such heavily used blocks map to the same set, then it is not possible to place all the blocks to the nearest bank. Hence, the power of migration in DNUCA is limited by the rules of set-associative cache. Many innovative techniques have been proposed to overcome this issue [12, 24]. In [12, 26], the author separates the tag and data part of the cache. The tag part is stored near to the core which requires very less time to search. The data are stored in the banks. The mapping between tag and data is maintained by forward and backward pointers. Once the tag is found, the corresponding forward pointer tells the location of the data. The traditional set-associative property is not required to maintain here as the mapping is done by forward/backward pointers. In this technique, more number of blocks can be migrated to the nearest bank.

All the above techniques proposed are applicable to both single-core and multicore processors (CMP). There are some unique issues with DNUCA to implement it for CMPs [1]. The main issue is migration; since a block can be requested by more than one cores, the block may every time move in different directions without stabilizing. In [10], the author proposed an improved migration rule which allows a block to migrate only after some fixed number of accesses. As the number of core increases, the size of LLC increases and also the number of banks. The bankset searching creates some new issues in CMPs. Now multiple cores can request for the same bank simultaneously as a result multiple search operations for the same block may be performed at the same time. Such simultaneous block search has to be handled carefully; otherwise, multiple miss request for the same block can be generated [10].

In [13], the author proposed a DNUCA-based CMP called HK-NUCA. In this design, the author maintains a home bank for each block. The home bank has the knowledge of where the bank is currently resides.

3.3 Data Replication

In CMP where multiple cores share the same LLC, sometimes it is beneficial to replicate some data. As DNUCA has a complex searching mechanism, researchers also try to improve SNUCA performance with data replication. In [27], the author proposed a technique to keep a copy of the heavily requesting blocks into the local bank (or nearby bank). On an L1 miss, the block will be first searched in its local bank, and if the block is not found in the local bank, then only the request is forwarded to the home bank (bank where the block maps). It reduces the latency to access the remote bank. But additional coherence mechanism must be maintained for the replication. Some other similar works proposed to improve SNUCA without using DNUCA are [6, 28].

4 Miss Rate

Reducing miss rate (*miss per instruction* or *miss per thousand instructions*) plays an important role for handling today's big data-based applications. Reduced miss rate means more hits in the cache and hence improves the system performance. As discussed in Sect. 1, using larger size cache is not the best technique to reduce miss rate. Today's LLC is already larger in size. Efficient use of such LLC can improve the miss rate even more.

In [8], the authors have observed that the entire space of the cache memory is not utilizing properly. In case of a set-associative cache, some sets are showing high miss rate, while some other sets are almost idle. Such non-uniform distribution reduces the utilization factor of the cache. Better cache utilization techniques can reduce the miss rate of the cache. Some of the work done to improve the utilization of LLC are [7–9, 19, 21, 29]. Most of these works are based on dynamic associativity management, i.e., increasing or decreasing the associativity of a set dynamically. The associativity of a heavily used set is increased by allowing the set to use some of the ways of other idle sets.

Another utilization issue is the non-uniform load distribution among the banks. Some banks are heavily loaded, while some other banks are almost idle [6, 10, 22, 23, 28]. Techniques have been proposed to evenly distribute the loads among all the banks. The most popular technique proposed is called cooperative caching [6]. In this technique, a heavily used bank can spill a block to its neighbor bank. Later the technique has been improved in [28]. Most of these techniques require an expensive centralized directory. In [10], the author proposed a DNUCA-based Tiled CMP to evenly distribute the loads among different banks.

Replacement policy and victim cache also play an important role in reducing miss rates [1]. In set-associative cache, each set has its own replacement policy. When a newly fetched block needs to be placed in cache and there is no space available for it, the replacement policy selects a victim block from the cache to replace it with the new block. Most of the traditional caches use least recently used (LRU) policy to select victim block from the cache. LRU policy is easy to implement, but it has some major issues like *dead block* and *never reused block*. A block is dead in cache if it will never be reused again before of its eviction. A never reused block accessed only once in cache. Such blocks unnecessary pollute the cache space. In LRU, they can only be evicted after becoming the least recently used block within the set. Techniques have been proposed to remove such dead blocks early [20, 30]. Using victim cache is also effective to reduce miss rate. In this technique, the evicted block is stored in a small storage called victim cache. If there is a miss in the cache, then the victim block is searched before going to main memory. A hit in victim cache prevents an off-chip main memory access. Some of the works done related to victim ache are [27, 31].

5 Near-Data Processing

The most innovative technique proposed for handling big data applications is called near-data processing (NDP) [32]. In the era of big data, all the application has a huge amount of data, and they also need to perform many expensive and time-consuming computations [33]. Such requirements result in developing specialized systems for high-performance computing. But even after proposing such high performance and parallelized computers, the cache memory and computation remain a bottleneck. Processing big data means executing some operation with such huge data. The operation can be searching, inserting, sorting, etc. It can also be some arithmetic operation like addition, subtraction, multiplication, division, bit operation, etc. SQL, written for handling huge data also, is an example of such computation. In the traditional computers, all such computations are done by the ALU of the cores. Hence, all the data must be loaded into the core for computation. As discussed above the CMP having multiple cores has complex internal structures, it has multiple cores, banks connected with on-chip network. It has been found that for today's applications the data communication cost is more dominant than the data computation cost [32]. The data communication cost means the cost of moving a block from memory to the core through all the level of caches. Such communication increases the network congestion, latency, and power consumption. Hence even after using many cores, the performance of the system may not be as expected.

Researchers have proposed an alternative technique called NDP to reduce such communication overhead [33–36]. In NDP instead of moving the data to cores for computation, the computation unit is moved closer to the memory [35]. To do this, some simple computational units (SCU) are placed near the memory units including main memory and LLC. The operations done by these SCUs are limited. For example, some SCU can only perform addition, while some other SCUs can perform comparison. Note that such SCUs are designed for a specific operation and cannot be programmed for general-purpose computation like cores [37]. For example, suppose a student name needs to be searched in a file of 8 GB. In NDP if there is a specialized SCU installed in memory, then the SCU can perform the computation in memory and the data is not required to send to the cores. Such near-data computation reduces the on-chip network overhead and hence improves the performance of LLC [37, 38].

6 Conclusion

For efficient execution of big data-based applications, the performance of the last-level cache (LLC) in today's high-performance computer must be enhanced. The multicore processors used in such computers placed multiple cores in a single chip called chip multiprocessor (CMP). Each core has its own private cache memories, and all the cores share a common LLC. Designing efficient LLC for CMP has many challenges like latency overhead, utilization overhead, etc. In this paper, we have done a survey on the innovative techniques proposed for the LLC of CMPs to efficiently handle big data-based applications.

References

1. R. Balasubramonian, N.P. Jouppi, N. Muralimanohar, *Multi-Core Cache Hierarchies* (Morgan and Claypool Publishers, 2011)
2. G.H. Loh, 3D-stacked memory architectures for multi-core processors. ACM SIGARCH Computer Architecture News **36**, 453–464 (2008)
3. J. Huh, C. Kim, H. Shafi, L. Zhang, D. Burger, S.W. Keckler, A NUCA substrate for flexible CMP cache sharing, in *Proceedings of the 19th Annual International Conference on Supercomputing (ICS)* (2005), pp. 31–40
4. U. Nawathe, M. Hassan, L. Warriner, K. Yen, B. Upputuri, D. Greenhill, A. Kumar, H. Park, An 8-Core 64-thread 64B power-efficient SPARC SoC, in *Proceedings of the IEEE International Solid-State Circuits Conference (ISSCC)* (2007), pp. 108–590
5. AMD Athlon 64 X2 Dual-Core Processor for Desktop. (Online). Available: http://www.amd.com/us-en/Processors/ProductInformation/0,,30118948513041,00.html
6. J. Chang, G.S. Sohi, Cooperative caching for chip multiprocessors, in *Proceedings of the International Symposium on Computer Architecture (ISCA)* (2006), pp. 264–276
7. S. Das, H.K. Kapoor, Dynamic associativity management using fellow sets, in *Proceedings of the 2013 International Symposium on Electronic System Design (ISED)* (2013), pp. 133–137
8. M.K. Qureshi, D. Thompson, Y.N. Patt, The V-way cache: demand based associativity via global replacement. ACM SIGARCH Comput Archit. News **33**(2), 544–555 (2005)
9. S. Das, H.K. Kapoor, Victim retention for reducing cache misses in tiled chip multiprocessors. Microprocess. Microsyst. **38**(4), 263–275 (2014)
10. S. Das, H.K. Kapoor, Exploration of migration and replacement policies for dynamic NUCA over tiled CMPs, in *Proceedings of the 28th International Conference on VLSI Design (VLSID)* (2015)
11. C. Kim, D. Burger, S.W. Keckler, An adaptive, non-uniform cache structure for wire-delay dominated on-chip caches. ACM SIGOPS Operating Syst. Rev. **36**, 211–222 (2002)
12. Z. Chishti, M.D. Powell, T.N. Vijaykumar, Distance associativity for high-performance energy-efficient non-uniform cache architectures, in *Proceedings of the 36th Annual IEEE/ACM International Symposium on Microarchitecture (MICRO)* (2003), pp. 55–66
13. J. Lira, C. Molina, A. Gonzalez, HK-NUCA: boosting data searches in dynamic non-uniform cache architectures for chip multiprocessors, in *Proceedings of the IEEE International Symposium on Parallel and Distributed Processing (IPDPS)*, (2011), pp. 419–430

14. W. Ding, M. Kandemir, Improving last level cache locality by integrating loop and data transformations, in *Proceedings of the IEEE/ACM International Conference on Computer-Aided Design (ICCAD)* (2012), pp. 65–72
15. J.L. Hennessy, D.A. Patterson, *Computer Architecture: A Quantitative Approach*, 4th edn. (Elsevier, 2007)
16. A. El-Moursy, F. Sibai, V-set cache design for LLC of multi-core processors, in *Proceedings of the IEEE 14th International Conference on High Performance Computing and Communication and IEEE 9th International Conference on Embedded Software and Systems (HPCC-ICESS)* (2012), pp. 995–1000
17. G.H. Loh, Y. Xie, B. Black, Processor design in 3D die-stacking technologies. IEEE Micro **27**(3), 31–48 (2007)
18. Intel. Quad-Core Intel Xeon Processor 5400 Series, apr 2008. (Online). Available: http://download.intel.com/design/xeon/datashts/318589.pdf
19. D. Rolán, B.B. Fraguela, R. Doallo, Adaptive line placement with the set balancing cache, in *Proceedings of the 42nd Annual IEEE/ACM International Symposium on Microarchitecture (MICRO)* (2009), pp. 529–540
20. S. Das, N. Polavarapu, P.D. Halwe, H.K. Kapoor, Random-LRU: a replacement policy for chip multiprocessors, in *Proceedings of the International Symposium on VLSI Design and Test (VDAT)* (2013)
21. D. Sanchez, C. Kozyrakis, The ZCache: decoupling ways and associativity, in *Proceedings of the 43rd Annual IEEE/ACM International Symposium on Microarchitecture (MICRO)* (2010), pp. 187–198
22. P. Foglia, M. Comparetti, A workload independent energy reduction strategy for D-NUCA caches. J. Supercomput. **68**, 157–182 (2013)
23. H. Kapoor, S. Das, S. Chakraborty, Static energy reduction by performance linked cache capacity management in Tiled CMPs, in *Proceedings of the 30th ACM/SIGAPP Symposium On Applied Computing (SAC)* (2015)
24. B.M. Beckmann, D.A. Wood, Managing wire delay in large chip-multiprocessor caches, in *Proceedings of the 37th Annual IEEE/ACM International Symposium on Microarchitecture (MICRO)* (2004), pp. 319–330
25. B.M. Beckmann, D.A. Wood, TLC: transmission line caches, in *Proceedings of the 36th Annual IEEE/ACM International Symposium on Microarchitecture (MICRO)* (2003), pp. 43–54
26. Z. Chishti, M.D. Powell, T.N. Vijaykumar, Optimizing replication, communication, and capacity allocation in CMPs. ACM SIGARCH Comput. Archit. News **33**, 357–368 (2005)
27. M. Zhang, K. Asanovic, Victim replication: maximizing capacity while hiding wire delay in tiled chip multiprocessors, *Proceedings of the 32nd Annual International Symposium on Computer Architecture (ISCA)*, vol. 0, (2005), , pp. 336–345
28. M. Hammoud, S. Cho, R. Melhem, Dynamic cache clustering for chip multiprocessors, in *Proceedings of the 23rd International Conference on Supercomputing (ICS)* (2009), pp. 56–67
29. D. Zhan, H. Jiang, S.C. Seth, STEM: spatiotemporal management of capacity for intra-core last level caches, in *Proceedings of the 2010 43rd Annual IEEE/ACM International Symposium on Microarchitecture (MICRO)* (2010), pp. 163–174
30. M.K. Qureshi, A. Jaleel, Y.N. Patt, S.C. Steely, J. Emer, Adaptive insertion policies for high performance caching. ACM SIGARCH Comput. Archit. News **35**(2), 381–391 (2007)
31. S.M. Khan, D.A. Jiménez, D. Burger, B. Falsafi, Using dead blocks as a virtual victim cache, in *Proceedings of the 19th International Conference on Parallel Architectures and Compilation Techniques (PACT)* (2010), pp. 489–500
32. R. Balasubramanian, J. Chang, T. Manning, J.H. Moreno, R. Murphy, R. Nair, S. Swanson, Near-data processing: Insights from a micro-46 workshop. IEEE Micro **34**(4), 36–42 (2014)
33. H. Asghari-Moghaddam, A. Farmahini-Farahani, K. Morrow, J.H. Ahn, N.S. Kim, Near-DRAM acceleration with single-ISA heterogeneous processing in standard memory Mmodules. IEEE Micro **36**(1), 24–34 (2016)
34. D. Park, J. Wang, Y.S. Kee, In-storage computing for Hadoop MapReduce framework: challenges and possibilities. IEEE Trans. Comput. **99**, 1–1 (2016)

35. S.H. Pugsley, A. Deb, R. Balasubramonian, F. Li, Fixed-function hardware sorting accelerators for near data MapReduce execution, in *Proceedings of the 33rd IEEE International Conference on Computer Design (ICCD)* (2015), pp. 439–442

36. B. Gu, A.S. Yoon, D.H. Bae, I. Jo, J. Lee, J. Yoon, J.U. Kang, M. Kwon, C. Yoon, S. Cho, J. Jeong, D. Chang, Biscuit: a framework for near-data processing of big data workloads, in *2016 ACM/IEEE 43rd Annual International Symposium on Computer Architecture (ISCA)* (2016), pp. 153–165

37. M. Gao, C. Kozyrakis, HRL: efficient and flexible reconfigurable logic for near-data processing, in *Proceedings of the IEEE International Symposium on High Performance Computer Architecture (HPCA)* (2016), pp. 126–137

38. V.T. Lee, A. Mazumdar, C.C.d. Mundo, A. Alaghi, L. Ceze, M. Oskin, POSTER: application-driven near-data processing for similarity search, in *2017 26th International Conference on Parallel Architectures and Compilation Techniques (PACT)* (2017), pp. 132–133

Dr. Hadoop Cures In-Memory Data Replication System

**Ripon Patgiri, Sabuzima Nayak, Dipayan Dev
and Samir Kumar Borgohain**

Abstract The replication system attracts many researchers to rethink of the tactics of data placement in the various devices, namely RAM and HDD/SSD. The advantages of a replication system are (a) parallelism, (b) data availability, (c) fault tolerance, (d) data recovery, and (e) failover. However, the replication system poses some overheads, namely, communication and synchronization cost. In this paper, we show an in-memory data replication system using the Dr. Hadoop framework. It adapts Dr. Hadoop framework for in-memory replication system. This, in turn, provides very high availability, scalability, and fault-tolerant nature for Dr. Hadoop metadata server. The paper delivers various theorems that are presented in different sections of the paper and are proved using theoretical analysis.

Keywords Hadoop · In-memory replication · Big Data · Fault tolerance · Failover · Data replication · high availability · Scalability · Dr. Hadoop

R. Patgiri (✉) · S. Nayak · S. K. Borgohain
National Institute of Technology Silchar, Silchar 788010, Assam, India
e-mail: ripon@cse.nits.ac.in
URL: http://cs.nits.ac.in/

S. Nayak
e-mail: sabuzimanayak@gmail.com

S. K. Borgohain
e-mail: samir@nits.ac.in
URL: http://cs.nits.ac.in/

D. Dev
Prompt Cloud Pvt. Ltd., Banglore, India
e-mail: dev.dipayan16@gmail.com

© Springer Nature Singapore Pte Ltd. 2020
A. Elçi et al. (eds.), *Smart Computing Paradigms: New Progresses and Challenges*,
Advances in Intelligent Systems and Computing 767,
https://doi.org/10.1007/978-981-13-9680-9_19

1 Introduction

The Big Data [5] is the most dominant paradigm nowadays in the IT industry. The Big Data is termed as an enormous amount of variety of data to store, process, and manage. There is explosive data growth in every field which makes difficult for a data scientist to analyze, visualize, and process [4, 10]. Moreover, most of the working set of the data is stored in HDD/SSD, and non-working set of data is stored in the Tape drive. The hotspot data are stored in the RAM. However, the hotspot data are enormous due to the explosive data growth in every field. In addition, the vendee of Big Data is also increasing [11].

In large-scale data-intensive computing, a huge explosion of data has to be handled efficiently and effectively. There are three ways of storing data as per the requirement of the storage system, namely RAM, HDD/SSD, and Tape. The primary storage is used for streaming purposes. Apache Spark [14] builds for data streaming using in-memory data storage system called RDD [13]. Moreover, Memcached is used caching object for faster response [6, 8]. Hyper constructs in-memory database system for OLTP and OLAP [7].

We assume the following key points:

- The cluster is formed using shared-nothing architecture.
- We consider the data as a black box. The data in the main memory are represented by HashMap, Table, Tree, Matrix, etc., and data on the HDD can be written in the file.
- We also assume the cluster with low-cost commodity hardware. The low-cost commodity hardware is unreliable.
- Finally, we consider the heterogeneity of hardware capability, but homogeneity for the size of RAM.

In this paper, we perform an analysis to adapt Dr. Hadoop for in-memory data replication system. The Dr. Hadoop is a well-suited framework for in-memory data replication system. In addition, Dr. Hadoop uses a minimum number of resources to achieve high availability by data replication, which is calculated as 99.99% uptime [3].

The paper is organized as follows—Sect. 2 discusses on Dr. Hadoop. Section 3 calculates the memory overhead. Section 4 explores the update mechanism on in-memory replication system. Section 5 exploits the communication strategy of Dr. Hadoop for in-memory data replication. Section 6 analyzes the capability of fault tolerance. Section 8 discusses various aspects of Dr. Hadoop in adapting for in-memory data replication system. Finally, the paper is concluded by Sect. 9.

2 Dr. Hadoop

The Hadoop Distributed File System (HDFS) is a replica of the Google File System (GFS). The HDFS is the most famous file system for large-scale data management. The HDFS works on low-cost commodity hardware. Therefore, the HDFS has extensively experimented the distributed file system. Moreover, the HDFS offers high availability, scalability, fine-grained fault tolerance, and supports unstructured data management. However, the HDFS faces a single point of failure (SPoF) [9, 12]. The HDFS comprises of a single active metadata server, known as NameNode. The NameNode serves metadata and does not store data. The NameNode is supported by a secondary NameNode which stores checkpoints of NameNode as FSImage. The FSImage is used to recover/restore the crashed/failed NameNode. But, the secondary NameNode does not act as hot standby. Besides, the HDFS suffers from small file problems. A large set of a small number of files consumes entire RAM spaces, and hence, the scalability of HDFS becomes questionable [1]. To address the above problem, the Dr. Hadoop is proposed by Dev and Patgiri [3]. The Dr. Hadoop overcomes SPoF, small file problem [2], load balancing issue, hotspot issue, and scalability issue [9, 12]. The Dr. Hadoop uses DCMS data structure which is based on a circular doubly linked list.

Dr. Hadoop is an in-memory replication design based on the Hadoop framework. The Dr. Hadoop starts with three nodes [3]. A node can be inserted into the Dr. Hadoop, and thus, it grows incrementally. The incremental scalability provides infinite scaling of data. Thus, the system grows as per the requirement of Dr. Hadoop. The Dr. Hadoop also supports the removal of the node, and thus, it also shrinks.

In most of the failover system, the data are kept in the idle node. On the contrary, a node of the Dr. Hadoop acts as an active server as well as standby for two other nodes. The Dr. Hadoop shares the load among three nodes. Therefore, the load of a node is always low naturally. The total number of nodes in the Dr. Hadoop is funded to the client.

2.1 Initialization

The Dr. Hadoop initializes with three nodes as shown in Fig. 1, and the minimum node requirement is also three nodes. Let us assume, the name of the nodes is A, B, and C. The nodes are arranged using the circular doubly linked list (CDLL) data structure. Therefore, the three nodes composed a CDLL. For example, $A \rightleftarrows B$, $B \rightleftarrows C$, and $C \rightleftarrows A$. The initial property follows the rule of CDLL, however, the CDLL can be initially empty. But, Dr. Hadoop cannot be empty, and it consists of three nodes. The RAM of node A contains its own data and also contains replicated data from node B and C. The RAM of node B contains its own data and also contains replicated data of node A and C. Similarly, the RAM of node C contains its own data and also contains replicated data of node A and B.

2.2 Insertion

Let us insert a node D between A and C. The insertion is exactly same with CDLL. The D is logically placed between A and C by recording the IP address of D. The

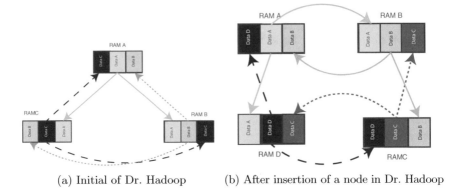

(a) Initial of Dr. Hadoop   (b) After insertion of a node in Dr. Hadoop

Fig. 1 Dr. Hadoop initialization and four nodes of Dr. Hadoop

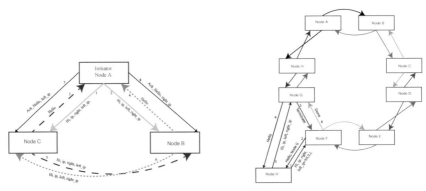

(a) Initial formation of three node cluster   (b) Insertion of a node in eight node cluster

Fig. 2 Insertion of nodes

node maintains the IP address of each node to know about the link structure. For example, the IP address list of node A as shown in Fig. 3 is

$$Right\ of\ A = IP\ address\ of\ B \rightarrow IP\ address\ of\ C \rightarrow \cdots$$

$$Left\ of\ A = IP\ address\ of\ H \rightarrow IP\ address\ of\ G \rightarrow \cdots$$

The RAM data of A and C are copied to RAM of D, and the RAM data of D are blank as shown in Fig. 2. That is, the RAM data of A and C are replicated in the

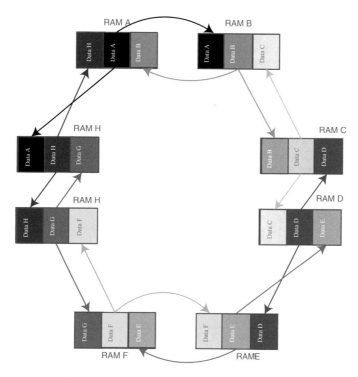

Fig. 3 In-memory replication design

RAM of *D*. The RAM data of *C* in RAM *A* are deleted, and then, created an empty space for *D* in RAM *A* for replication of data *D*. Similarly, the RAM data of *A* in RAM *C* are deleted, and then, created an empty space for *D* and RAM *D* replica is saved in RAM *C*.

2.3 Deletion

The deletion operation causes deletion of data in other nodes too. For example, the removal of node *D* causes readjustment in Fig. 2 to form the structure as Fig. 1. The RAM data of *D* and replicated data of *D* are deleted. Replace the RAM data of *D* in RAM *A* by copying the RAM data of *C*. Similarly, replace the RAM data of *D* in RAM *C* by copying the RAM data of *A*. Now, remove the node *D*.

3 Memory Space Requirements

Theorem 1 *Each node of a Dr. Hadoop cluster contributes $\frac{m}{3}$ to the cluster where m is the least size of available memory space of the system.*

Proof Let n the total number of nodes in a Dr. Hadoop cluster. Given that m is the least size of available memory space of the system. The Dr. Hadoop framework uses three nodes at the very beginning of the cluster. Let A be a node that shares the memory space with two neighbors, say node B and C. The available memory space of A is shared with the two nodes, and thus, the available memory space of node A is $\frac{m}{3}$. The remaining available memory spaces of node A is used by node B and C. Therefore, total available memory space for the cluster using node A, B, and C is

$$= \frac{m}{3} + \frac{m}{3} + \frac{m}{3} = m$$

The total memory capacity of the three nodes is m where each node's RAM capacity is m. Now, four nodes capacity is calculated as follows

$$= \frac{m}{3} + \frac{m}{3} + \frac{m}{3} + \frac{m}{3}$$

$$= m + \frac{m}{3}$$

Therefore, the accumulated memory space of n nodes is

$$= \frac{m}{3} + \frac{m}{3} + \frac{m}{3} + \frac{m}{3} + \frac{m}{3} + \cdots \ up\ to\ n$$

$$= \frac{mn}{3}$$

For example, if each node contains 6 GB of RAM, then the total capacity of twenty-four (24) nodes is $24 \times \frac{6}{3}$ GB $= 48$ GB. Therefore, each node contributes $\frac{m}{3}$ memory spaces if the node contains m memory spaces.

4 Update Replicated Data

The data are replicated in two neighbors of a node, called *Left* and *Right*. The replicated data are updated when data of a node are updated. Modification of data can happen at any time, and it may be too frequent or too lazy. However, the number of data modification depends on the situation. Immediate update on the replicated node on modification can slow down the system. Because there may be thousands of updates per second on a system which triggers update immediately on the replicated node. It may lead to thousands of performances of network communication in a second. Therefore, update too eagerly on replicated data is avoided in the general system. On the contrary, it is very important in mission-critical system. Therefore, the time *threshold* is set to zero for mission-critical system or any important system. The *threshold* signifies times to sleep, a flag *UFlag* is used to signify whether a modification has been done or not, and a log *ULog* contains the updated data to send to its neighbors. The *UFlag* is a volatile variable to keep track of update status. The *Event* is event manager, and it becomes true on modification of data.

5 Communication

5.1 Permanent Storage

The data are kept in RAM as well as flushed into HDD/SSD periodically. The data are written back to the HDD/SSD periodically depending on the threshold value. If the threshold value is zero, then update on data triggers write back to HDD/SSD. The replicated data are also kept in RAM as well as HDD/SSD. The writing into HDD/SSD is an integral part of Dr. Hadoop since data are placed in RAM.

5.2 The Heartbeat Protocol

The Dr. Hadoop inherits the property of the heartbeat mechanism of HDFS. The liveliness of a node is detected using heartbeat signal. If a node does not receive a heartbeat signal from its neighbor for a certain time, for example, 5 s, then the node is assumed to be dead. A node sends two heartbeat signals to the left node and right node. Thus, Dr. Hadoop ensures the liveliness of a node. Let X be the failed node, W be the left node of X, and Y be the right node of X. The W and the Y of the X merge (Link) and continue to serve until a new node is inserted. On this case, the W and Y contain single hot-standby node, i.e., the left node of W and right node Y will be the hot standby. Moreover, the data of node X are retained by W and Y node.

Lemma 1 *Dr. Hadoop encourages three parallel read operations, but not write operation.*

The read operation can be performed in three parallel nodes. For example, data of node *A* can be read from node *H* and node *B*. Any update (write) operation requires to lock the data on node *H* and node *B*, then reflects the data to the node *H* and node *B*. The reflection of modification data depends on the consistency model. Loose consistency and strict consistency can be applied in the system for general system and mission-critical system.

6   Fault Tolerance

The fault tolerance is one of the most prominent reasons to design a replicated model. Let us assume, the Dr. Hadoop is deployed in low-cost commodity hardware. Thus, machines are prone to fail. Any machine can fail at anytime and anywhere. The cases for faults in the machine are enlisted below:

– Machine crashes
– Network failure.

However, the above two cases cannot be differentiated. There is no way to know which occurred, whether a machine failure or a network failure. The heartbeat is sent to each other to inform their liveliness. For example, node A sends its heartbeat signal to node B and node H periodically (5 second interval) as shown in Fig. 3. If the heartbeat signal is not received from node H, then the node A assumes that node H is down after a period of time, say 5 min.

6.1   Machine Crashes

Let us assume, some nodes fail one by one. Let us assume node A is dead, and thus, the node H and node B contains the data of node A in main memory referring to Fig. 4. If node B fails, then the node C contains the data of the node B and node A in main memory and HDD, respectively. Because the node C cannot share its main memory space to more than two. Now, node C is also dead. Then, the node D contains the data of node C in main memory. The node D contains data of node A and node B in HDD/SSD for recovery purposes. In this case, the node H also stores current failure data in RAM. Other data of failure nodes are stored in HDD/SSD. Thus, Dr. Hadoop ensures data lossless.

Theorem 2 *Dr. Hadoop can survive a single node failure without affecting the system.*

Proof Let us assume, *Left* of node A is node H and *Right* of A is node B referring to Fig. 3. The node B and H have not heard any heartbeat signal from node A for some period of time referring to Fig. 4. The Case I denotes that the node A is down in Fig. 4. Therefore, the nodes B and H add the node A in the failure list, and then, communicates between node B and H through query message to know the status. The nodes B and H confirm the failure of node A. Now, the node A is unable to serve. Therefore, nodes B and H serve on behalf of failure node A. If the threshold is not equal to zero, then all recent transactions from the last checkpoint have been lost. Therefore, the node B and node H serve the data of A from the last checkpoint. Thus, the Dr. Hadoop can tolerate single machine failure.

Theorem 3 *The Dr. Hadoop can survive at most two consecutive node failure at a given time T_i without affecting the system.*

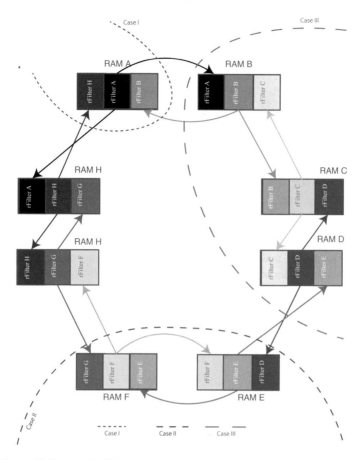

Fig. 4 Cases of failures in Dr. Hadoop

Proof Let us assume, the node A and node B are down for some period at a given time referring to Fig. 4. The Case II denotes the failure of node A and node B in a given time as shown in Fig. 4, i.e., two consecutive node failure at a given time. The node H has not heard the heartbeat signal for some period of time. Therefore, the node H tries to communicate B through a query message where there is no response from node B (may be packet drop or crashes). Immediately, node H communicates node C where node C responses. If node C has a similar experience on node B and node A, then both node A and node B are dead. Now, node H and node C can update the failure list. In this case, the data of node A is served by node H and the data of node B is served by node C. Thus, the Dr. Hadoop can tolerate two consecutive machine failures at a given time.

Theorem 4 *Dr. Hadoop loses single node data in three consecutive node failure at a given time T_i.*

Proof Let us assume the node *A*, node *B*, and node *C* are down for some period at a given time referring to Fig. 4. The Case III denotes the failure of node *A*, node *B*, and node *C* at a given time as depicted in Fig. 4, i.e., three consecutive node failures at a given time. In this case, node *H* has not heard the heartbeat signal from node *A* for some period of time. It is also similar to the above-cited example of two consecutive node failures at a given time, except three consecutive node failures at a given time. The data of node *A* are served by node *H*, and the data of node *C* are served by node *D*. There is no node to serve the data of node *B*. Therefore, the Dr. Hadoop cannot tolerate the three consecutive node failures at a given time.

Theorem 5 *The Dr. Hadoop can survive multiple non-consecutive node failures at a given time.*

Proof The Dr. Hadoop tolerates many numbers of non-consecutive node failure at a given time. Let us assume node *A*, nodes *C*, and *E* fail at a given time referring to Fig. 3. The data of node *A* are served by node *H*, the data of node *C* are served by node *B*, and the data of node *E* are served by node *D*. Thus, Dr. Hadoop tolerate multiple non-consecutive node failures at a given time.

Let node *C* is dead after some time. The node *A* contains the data of node *C* in RAM. However, the node *A* contains the data of node *B* in HDD/SSD, but node *A* cannot serve the client until a new node is inserted into the system. Thus, Dr. Hadoop ensures the availability of multiple non-consecutive failures.

Theorem 6 *Dr. Hadoop can survive any number of two consecutive node failure at a different time.*

Proof Dr. Hadoop can tolerate multiple non-consecutive node failures at a given time. Let us assume node *A* is down. Now, another node is inserted to replace the failed node *A*. Therefore, the node *A* is alive after some time. After that, node *B* and then node *C* are down one after another. Node *B* can be recovered by replacing using another node. Then, the failure of node *C* is not a problem at all. Node *C* can also be recovered by replacing using another node. The insertion of nodes makes the Dr. Hadoop unique and robust.

6.2 Network Failure

Let us assume the connection between the node *H* and node *A* is broken referring to Fig. 3. Now, the node *H* has not heard the heartbeat signal from node *A* for a long period of time, whereas the node *B* has heard recently and quite regularly. Then, the node *H* communicates with node *B*, but the node *B* informs the liveliness of node *A* to node *H*. Also, the node *B* informs the incident to node *A*. Both node *A* and node *H* assume that there is a link failure between them.

7 Mean Time Between Failure

Let, $R(t)$ be reliability function of a node at time $[0, t]$, and μ be the average failure rate. The reliability function is expressed using the Poisson distribution function which is

$$R(t) = e^{-\mu t}$$

$R(t)$ is defined no machine failure at time $[0, t]$. The probability of failure of a node is

$$P(t) = 1 - R(t)$$
$$= 1 - e^{-\mu t}$$

Thus, we can derive the probability density function (for failure) as

$$f(t) = \frac{dP(t)}{dt}$$
$$= \mu e^{-\mu t}$$

7.1 Single Node

The probability of failure of a node between time $[t_0, t_1]$ is

$$P(t_0 \rightarrow t_1) = \int_{t_0}^{t_1} f(t) dt$$
$$= \mu \int_{t_0}^{t_1} e^{-\mu t} dt$$
$$= e^{-\mu t_0} - e^{-\mu t_1}$$

Therefore, the mean time between failure for a single node n is calculated as

$$mtbf(n) = \int_{0}^{\infty} t f(t) dt$$
$$= \frac{1}{\mu}$$

7.2   Two Consecutive Nodes' Failure

Mean time between failure of two consecutive nodes is $mtbf(n_1; n_2)$. Thus,

$$mtbf(n_1; n_2) = \frac{1}{\frac{1}{mtbf(n_1)} + \frac{1}{mtbf(n_2)}}$$
$$= \frac{mtbf(n_1) \times mtbf(n_2)}{mtbf(n_1) + mtbf(n_2)}$$

7.3   Three Consecutive Nodes' Failure

Similar to two consecutive nodes' failure, three consecutive nodes' failure also depend on each other, and thus,

$$mtbf(n_1; n_2; n_3) = \frac{1}{\frac{1}{mtbf(n_1)} + \frac{1}{mtbf(n_2)} + \frac{1}{mtbf(n_3)}}$$
$$= \frac{mtbf(n_1) \times mtbf(n_2) \times mtbf(n_3)}{mtbf(n_1) \times mtbf(n_2) + mtbf(n_2) \times mtbf(n_3) + mtbf(n_1) \times mtbf(n_3)}$$

However, the non-consecutive node failure does not depend on each other.

8   Discussion

The Dr. Hadoop network topology is logical ring-based topology, but underlying physical network topology can be a star, ring, mesh, bus, and hybrid topology. A rack-aware network topology can also be constructed similar to rack-aware data replication. The rack-aware network topology selects two nodes from one rack and one node from other racks. Therefore, it can tolerate rack failure too. The applications of Dr. Hadoop, as in-memory data replication system, are listed below

– A very large-scale Bloom filter.
– Storing large-scale Key/Value (e.g., HashMap).
– Storing large-scale metadata.
– In-memory databases.

However, the in-memory data replication is expensive. Besides, Dr. Hadoop is inapplicable in frequent data changes, for instance, web caching.

9 Conclusion

An in-memory data replication system plays a vital role in infrequent data changing. The Dr. Hadoop is adapted for in-memory data replication with uptime of 99.99%. Moreover, the Dr. Hadoop uses the least resources to achieve hot-standby and replication system. It also enables parallel read operation data. In addition, the load of a server is also balanced, and each node shares their loads with its left node and right node. Thus, the load of a node becomes one-third of a node's load. A node serves an active node as well as the passive node for two other nodes. Dr. Hadoop uses one phase commit protocol to reduce network congestion. In addition, the rack-aware network topology can also tolerate rack failure. Besides, the Dr. Hadoop can tolerate two consecutive node failure at a given time. Dr. Hadoop can tolerate many non-consecutive node failures at a given time or different time. Therefore, the Dr. Hadoop is the best choice to design an in-memory data replication system.

References

1. D. Dev, R. Patgiri, Performance evaluation of HDFs in big data management, in *2014 International Conference on High Performance Computing and Applications (ICHPCA)* (2014), pp. 1–7
2. D. Dev, R. Patgiri, HAR+: archive and metadata distribution! why not both? in *2015 International Conference on Computer Communication and Informatics (ICCCI)* (2015), pp. 1–6
3. D. Dev, R. Patgiri, Dr. Hadoop: an infinite scalable metadata management for Hadoop—how the baby elephant becomes immortal. Front. Inf. Technol. Electron. Eng. **17**(1), 15–31 (2016)
4. D. Dev, R. Patgiri, A survey of different technologies and recent challenges of big data, in *Proceedings of 3rd International Conference on Advanced Computing, Networking and Informatics*, ed. by A. Nagar, D.P. Mohapatra, N. Chaki (Springer, New Delhi, India, 2016), pp. 537–548
5. D. Dev, R. Patgiri, A survey of different technologies and recent challenges of big data, in *Proceedings of 3rd International Conference on Advanced Computing, Networking and Informatics* (2015), pp. 537–548
6. B. Fitzpatrick, Distributed caching with memcached. Linux J. **2004**(124), 5 (2004)
7. A. Kemper, T. Neumann, Hyper: a hybrid OLTP & OLAP main memory database system based on virtual memory snapshots, in *2011 IEEE 27th International Conference on Data Engineering* (2011), pp. 195–206
8. R. Nishtala, H. Fugal, S. Grimm, M. Kwiatkowski, H. Lee, H.C. Li, R. McElroy, M. Paleczny, D. Peek, P. Saab et al., Scaling memcache at facebook. in *NSDI*, vol. 13 (2013), pp. 385–398
9. R. Patgiri, Mds: in-depth insight, in *2016 International Conference on Information Technology (ICIT)* (2016), pp. 193–199
10. R. Patgiri, Issues and challenges in big data: a survey, in *Distributed Computing and Internet Technology*, ed. by A. Negi, R. Bhatnagar, L. Parida (Springer, Cham, 2018), pp. 295–300
11. R. Patgiri, A. Ahmed, Big data: the v's of the game changer paradigm, in *2016 IEEE 18th International Conference on High Performance Computing and Communications; IEEE 14th International Conference on Smart City; IEEE 2nd International Conference on Data Science and Systems (HPCC/SmartCity/DSS)* (2016), pp. 17–24
12. R. Patgiri, D. Dev, A. Ahmed, dMDS: uncover the hidden issues of metadata server design, in *Progress in Intelligent Computing Techniques: Theory, Practice, and Applications*, ed. by P.K. Sa, M.N. Sahoo, M. Murugappan, Y. Wu, B. Majhi (Springer, Singapore, 2018), pp. 531–541

13. M. Zaharia, M. Chowdhury, T. Das, A. Dave, J. Ma, M. McCauley, M.J. Franklin, S. Shenker, I. Stoica, Resilient distributed datasets: a fault-tolerant abstraction for in-memory cluster computing, in *Proceedings of the 9th USENIX Conference on Networked Systems Design and Implementation* (USENIX Association, Berkeley, CA, USA, 2012), pp. 2–2
14. M. Zaharia, M. Chowdhury, M.J. Franklin, S. Shenker, I. Stoica, Spark: cluster computing with working sets. HotCloud **10**(10–10), 95 (2010)

Communication Systems, Antenna Research, and Cognitive Radio

An Optimized OpenMP-Based Genetic Algorithm Solution to Vehicle Routing Problem

Rahul Saxena, Monika Jain, Karan Malhotra and Karan D. Vasa

Abstract Vehicle routing problem is an interesting combinatoric research problem of NP-complete class for investigation. Many researchers in the past have targeted this interesting combinatorial problem with a number of methodologies. The classic methods like brute-force approach, dynamic programming, and integer linear programming methods were used in earlier attempts to find the most optimized route for a vehicle. However, these methods met their computational limitation for a large number of coverage points. Owing to the exhaustive evaluation for the number of routes, the genetic algorithm-based heuristic approach was proposed to find accurate approximate solutions. The method involves solving a traveling salesman problem (TSP) using the genetic algorithm approach for a large number of route combinations which were very high. This research document proposes a solution to this by using the multi-core architecture, where it has been shown that implementing GA as heuristic approach for a large solution space is not sufficient. A contrast has been shown between the serial and parallel implementation of the solution using OpenMP multi-processing architecture which shows a considerable speedup for the execution time of the algorithm to search the best path. For a varied degree of graph structures, this implementation has highly reduced execution time.

Keywords NP-complete · Approximation · TSP · OpenMP · Multi-core

R. Saxena (✉) · M. Jain · K. Malhotra · K. D. Vasa
School of Computing and Information Technology, Manipal University Jaipur, Jaipur, India
e-mail: rahulsaxena0812@gmail.com

M. Jain
e-mail: monikalnct@gmail.com

K. Malhotra
e-mail: kmalhotra30@gmail.com

K. D. Vasa
e-mail: starkdv123@gmail.com

© Springer Nature Singapore Pte Ltd. 2020
A. Elçi et al. (eds.), *Smart Computing Paradigms: New Progresses and Challenges*,
Advances in Intelligent Systems and Computing 767,
https://doi.org/10.1007/978-981-13-9680-9_20

237

Fig. 1 Vehicle routing
algorithm

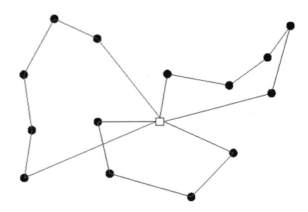

1 Introduction

Traveling salesman problem (TSP) is defined as a combinational structure problem
to find the most optimal path for a graphical structure mapped to the problem having
"n" vertices with a unique visit to each vertex. TSP finds its applicability in many real-
world problems like DNA sequencing and electrical network design [1]. The problem
due to its wide usage and applicability in optimization has caught the attention to
bring optimality over the solutions to some of the most computationally complex
problems. This paper focuses on the optimization of a well-known vehicle routing
problem (VRP).

The problem was first brought into notice by George Danzig and John Ramser in
a paper in 1959. The main depots, customers, vehicles, and routes being the main
components, the problem is defined as fulfilling the customer demands in a way
that each station is visited exactly once and the route with minimum cost covers all
the stations/depots. This problem falls under the category of NP-hard and similarizes
TSP. Figure 1 shows a graphical representation of the problem, where "n" stations are
shown and the route covers all the nodes starting from a source vertex to a destination.

2 Genetic Algorithm

The genetic algorithm is a natural selection process of the fit population/solutions.
The best fit solutions for each generation, generally 10–15% known as chromosomes
are retained, while discarding the remaining.

The genetic algorithm has certain points of difference in the solution imparting
procedure from the traditional methods:

- GA searches parallelly for solutions instead of a single point of attack.
- Selections are based on probabilistic measures rather deterministic.

- Solutions are obtained over the encoded population, i.e., not in terms of the actual input.
- Fitness scores are determined based on an objective function rather than having a derivative or other auxiliary information.

Genetic Algorithm Steps

- Create initial population P.
- Evaluate the fitness of each chromosome.
- Select Q parents from the current population.
- Choose a pair of parents for mating.
- Repeat Step 4 until all parents are selected and mated.
- Find fit chromosomes based on the fitness function.
- Repeat steps 3–6 for successive iterations until constant results are obtained.

The above algorithm helps to evaluate the most optimal route over a graphical structure developed for the problem.

3   Related Work and Studies

- Lin et al. in [2] discussed a mechanism for a shared ride with effective resource utilization and minimized traffic congestion.
- Nazif et al. [3] showed an optimal crossover technique to find the solution.
- Masum et al. [4] used a balanced technique to approach a solution quicker by reducing the search space.
- Chand et al. [5] solved multi-objective VRP using dominant ranking method.
- Kang and Jung [6] proposed a solution to multi-depot problem.
- Cooray [7] used a machine learning-based parameter-tuned GA for energy minimization.
- Nazif et al. [8] discussed a time window-constrained approach genetic algorithmic solution for solving using an optimized crossover operator.
- Alba et al. [9] discussed the cellular genetic algorithm approach for VRP.

From the above discussion, it can be concluded that VRP has been in investigation with various strategies to get more accurate and rapid solutions. In this paper, a parallelized version of GA-based solution to VRP has been shown.

4 Parallelization of Vehicle Routing Problem with OpenMP

To enhance the computational evaluation procedure for VRP, OpenMP programming provides the parallelization scope for the problem. OpenMP enables parallel code development to map to the multi-core architecture. OpenMP has three basic components: directives, environmental variables, and clauses. Multiple threads are generated as shown in Fig. 1, and work sharing is done between the threads using (i) directives, (ii) clauses, and (iii) environmental variables. OpenMP program contains both serial and parallel portions, where compiler directives [10] state which part is serial and which part is parallel. The flavor of parallel processing using OpenMP platform greatly enhances the performance and reduces the computational burden [10–13]. OpenMP supports C and FORTRAN languages. Parallel implementation of the algorithm is done over a varied number of cores and threads [14], and comparative analysis over the results has been done in this paper. Figure 2 showcases the basic OpenMP model.

Single-Core and Multi-Core Architecture
Single-core architecture uses one CPU and its memory to compute the problem as shown in Fig. 3 while in multi-core architecture quad-core (four processors are used) with shared memory and their own individual memory.

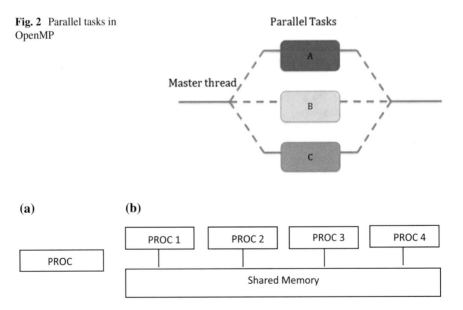

Fig. 2 Parallel tasks in OpenMP

Fig. 3 a Single-core architecture; **b** multi-core architecture

5   Implementation and Methodology

As vehicle routing protocol is a complete graph, the total number of ways to find possible paths is $n!$ As compared to the traditional method, the genetic algorithm increases the probability of finding the optimal path more rapidly. OpenMP has been incorporated to reduce the computation time of VRP.

To find the optimal solution, the steps followed are

Initialization: The initial set of chromosomes constitutes the initial possible routes with optimal cost [15]. In this step, randomly generated population is taken. Initial population should be generated carefully to fasten the genetic algorithm process.

Selection: Chromosomes having route cost lesser than the threshold calculated as per the fitness function selected are retained.

Fitness Function: The objective function for the evaluation of the fit population is given as follows

$$\textbf{Fitness Function} = (\textbf{max\_route\_cost} + \textbf{min\_route\_cost})/\textbf{2}$$

max_route_cost indicates the maximum value for route cost of all the available choices, and min_route_cost defines the lowest route cost in the current population. Parallelization of fitness function has been done by OpenMP.

Reproduction: Crossover and mutation form the basis of regenerating new chromosomes from the parents.

Chromosome: Each individual chromosome represents a possible tour from source to destination.

If there are five cities, then the possible chromosomes are

- {10, 13, 14, 11, 12}
- {13, 14, 10, 12, 11}
- {11, 12, 13, 14, 10} | and so on …

A total of $N!$ chromosomes are possible, where N represents the number of cities in consideration. Each chromosome represents an individual which is one possible solution to the vehicle routing problem.

Crossover: Crossover methodology has been used to generate new offsprings. Crossover operator is applied between the chromosomes to create new chromosomes. To restrict infinite growth, an upper limit holds for a number of crossovers. Parallelization function has been used in crossover function as well. The crossover technique used by us has been referred from the "Annals of Operations Research" Journal. To be more specific, we have used "Ordinal Representation" of Chromosomes.

6   Scope of Parallelization

We have executed the vehicle routing problem in multi-core architecture instead of single-core architecture. OpenMP is used for running the problem parallelly [16].

For generating initial populations over n cities, the algorithm is as follows:

Initial Population Parallelization over OpenMP

Step 1: Assigning a directive to instruct the compiler to run the code parallelly with n threads.
Running steps 2–4 over n threads using pragmas
Step 2: Set the minimum path value randomly.
Step 3: Compare the initial route with the current node.
Step 4: Set the minimum route from the initial and current route.
Step 5: Repeat Step 2, for n number of cities.

Chromosomes having route cost lesser than the threshold survive, and the rest are eliminated. The algorithm used for finding the current nodes is as follows:

Fitness Function Parallelization over OpenMP

Step 1: Assigning a directive to instruct the compiler to run the code parallelly with n threads.
Running steps 2–4 over n threads using pragmas
Step 2: Compare if calculated route cost is less than (min_route_cost + max_route_cost)/2.
Step 3: Reserve this iteration route in possible route vector.
Step 4: Repeat Step 2, for initial routes and n number of cities.

Fitness function and crossover are used for parallelization by OpenMP. We have used Thread building block library Intel TBB as it contains concurrent vector which is thread safe. Reduction function was used for parallelization.

Parallelization of **Ordinal Crossover** over OpenMP

Step 1: Assigning a directive to instruct the compiler to run the code parallelly with n threads.
Running steps 2–4 over n threads using pragmas
Step 2: For every chromosome, apply crossover technique.
Step 3: Reserve in vector child.
Step 4: Repeat Step 2, for every chromosome.

The above techniques are repeated until we find the optimal solution or once the cost is stable.

Experimental Results
Here, the genetic algorithm-based VRP over single-core and multi-core architecture

has been implemented and the two implementation strategies have been compared. Single-core or serial implementation although performs well with respect to the traditional VRP solutions discussed in terms of convergence and reduced execution timing but the parallel implementation over multi-core architecture using OpenMP programming model supersedes all. The results obtained are given in Table 1. As the number of cities or problem size grows exponentially, the execution timing difference between the serial and parallel implementation increases.

The above discussion shows that the time taken by multi-core version-based VRP implementation is nearly half as of the single-core implementation. Nearly, 50% speedup is obtained for parallel version. However, process iterations and optimal solution accuracy govern the speedup obtained. Graph 1 depicts the speedup obtained for a different number of nodes of multi-core-based implementation over serial. X-axis represents the various numbers of routes, and Y-axis shows the execution timing in seconds.

7 Conclusion

In this paper, we have discussed the special case of TSP. Approaching VRP using GA has been discussed in detail. When vehicle routing problem was executed over multi-core architecture, significant speedup in running time of the program is observed. However, the initial population and population generated determine the exact speedup achieved. In the future [14], different crossover techniques for quicker and more approximate results can be looked for. Moreover, some specialized GA techniques like non-dominant sorting-based GA (NSGA) and their parallel versions can be investigated to get more optimized results in quick time.

Table 1 Single-core versus multi-core architecture execution timing results for VRP

S. No.	Number of graph nodes	Single-core execution timing(s)	Multi-core execution timing(s)
1	5	4.620	1.227
2	7	8.859	9.973
3	10	17.82	7.245
4	15	26.498	11.047
5	20	14.334	8.10
6	25	26.282	9.668
7	50	34.567	16.914
8	60	30.147	18.965

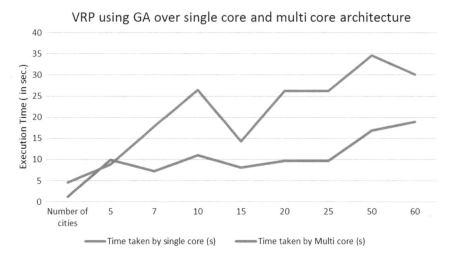

Graph 1 Single-core versus multi-core GA-based implementation of VRP

References

1. R. Saxena, M. Jain, D.P. Sharma, S. Jaidka, A review on VANET routing protocols and proposing a parallelized genetic algorithm based heuristic modification to mobicast routing for real time message passing. J. Intell. Fuzzy Syst. **36**(3), 2387–2398 (2019)
2. Y. Lin, W. Li, F. Qiu, H. Xu, Research on optimization of vehicle routing problem for ride-sharing taxi. Procedia Soc. Behav. Sci. **43**, 494–502 (2012)
3. H. Nazif, L.S. Lee, Optimized crossover genetic algorithm for capacitated vehicle routing problem. Appl. Math. Model. **36**(5), 2110–2117 (2012)
4. A.K.M. Masum, M. Shah Jalal, F. Faruque, I.H. Sarker, Solving the vehicle routing problem using genetic algorithm. Int. J. Adv. Comput. Sci. Appl. **2**(7), 126–131 (2011)
5. P. Chand, J.R. Mohanty, A multi-objective vehicle routing problem using dominant rank method. Int. J. Comput. Appl. 29–34 (2013)
6. R.G. Kang, C.Y. Jung, The improved initialization method of genetic algorithm for solving the optimization problem, in *International Conference on Neural Information Processing* (Springer, Berlin, Heidelberg, 2006), pp. 789–796
7. P.L.N.U. Cooray, T.D. Rupasinghe, Machine learning-based parameter tuned genetic algorithm for energy minimizing vehicle routing problem. J. Ind. Eng. (2017)
8. H. Nazif, L.S. Lee, Optimized crossover genetic algorithm for vehicle routing problem with time windows. Am. J. Appl. Sci. **7**(1), 95 (2010)
9. E. Alba, B. Dorronsoro, Solving the vehicle routing problem by using cellular genetic algorithms, in *European Conference on Evolutionary Computation in Combinatorial Optimization* (Springer, Berlin, Heidelberg, 2004), pp. 11–20
10. M. Jain, R. Saxena, V. Agarwal, A. Srivastava, An OpenMP-based algorithmic optimization for congestion control of network traffic, in *Information and Decision Sciences* (Springer, Singapore, 2018), pp. 49–58
11. R. Saxena, M. Jain, D. Singh, A. Kushwah, An enhanced parallel version of RSA public key crypto based algorithm using OpenMP, in *Proceedings of the 10th International Conference on Security of Information and Networks* (ACM, 2017), pp. 37–42
12. R. Saxena, M. Jain, D.P. Sharma, GPU-based parallelization of topological sorting, in *Proceedings of First International Conference on Smart System, Innovations and Computing* (Springer, Singapore, 2018), pp. 411–421

13. M. Jain, R. Saxena, Parallelization of video summarization over multi-core processors. Int. J. Pure Appl. Math. **118**(9), 571–584 (2018). ISSN: 1311-8080
14. R. Saxena, M. Jain, A. Kumar, V. Jain, T. Sadana, S. Jaidka, An improved genetic algorithm based solution to vehicle routing problem over OpenMP with load consideration, in *Advances in Communication, Devices and Networking* (Springer, Singapore, 2019), pp. 285–296
15. M. Basthikodi, W. Ahmed, Parallel algorithm performance analysis using OpenMP for multi-core machines. Int. J. Adv. Comput. Technol. (IJACT) **4**(5), 28–32 (2015)
16. R. Saxena, M. Jain, S. Bhadri, S. Khemka, Parallelizing GA based heuristic approach for TSP over CUDA and OPENMP, in *2017 International Conference on Advances in Computing, Communications and Informatics (ICACCI)* (IEEE, 2017), pp. 1934–1940

A Novel Computationally Bounded Oblivious Transfer with Almost Non-trivial Communication

Radhakrishna Bhat and N. R. Sunitha

Abstract Existing privacy preserving schemes have attained maximum privacy level either using the exponential modular operations or the strong intractability assumptions. But they failed to guarantee the following challenges till date namely (1) computationally bounded single database scheme with non-trivial communication (2) inbuilt integrity support (3) only linear encryption operations. We have proposed a single database Oblivious Transfer (OT) or Symmetric Private Information Retrieval (SPIR) schemes using the quadratic residuosity as the underlying cryptographic primitive. In this paper, we have constructed a new quadratic residuosity based concurrently executing recursive 2-bit encryption function which fulfill all the above mentioned challenges. This recursive 2-bit encryption functions receive Quadratic Residuosity Assumption (QRA) based queries and produce the reasonable communication bits as the response bits. The greatest advantages of the proposed schemes is that they operate on the plain database and the concurrently executing recursive 2-bit encryption functions involve only linear number of modular multiplications.

Keywords Oblivious transfer · Symmetric private information retrieval · Non-trivial communication · Computationally bounded · Quadratic residuosity assumption · Quadratic residuosity

1 Introduction

Consider a scenario: *user* wishes to retrieve h specified messages from n message database from the *database server* with a non-trivial communication cost in such a

R. Bhat (✉) · N. R. Sunitha
Department of Computer Science and Engineering, Siddaganga Institute of Technology, Affiliated to VTU Belagavi,
B H Road, Tumakuru 572 103, Karnataka, India
e-mail: rsb567@gmail.com

N. R. Sunitha
e-mail: nrsunithasit@gmail.com

© Springer Nature Singapore Pte Ltd. 2020
A. Elçi et al. (eds.), *Smart Computing Paradigms: New Progresses and Challenges*,
Advances in Intelligent Systems and Computing 767,
https://doi.org/10.1007/978-981-13-9680-9_21

way that the *server* doesn't learn anything about the messages retrieved and the *user* doesn't get any other message other than the specified messages.

Private Information Retrieval (PIR) is one of the *user privacy* preserving protocols in which the user can retrieve $h = 1$ message bit from n message bits database but there is no guarantee that the user retrieves only h messages. Private Block Retrieval (PBR) is the extension of PIR in which each message is treated as a block of bits instead of a single bit. Therefore, this concept only preserves *user privacy*. Hence, above scenario cannot be fulfilled with PIR alone.

It is possible to solve the above scenario using the secure communication concept of Oblivious Transfer (OT). The concept of oblivious transfer was introduced by Rabin [1] which is the basic building block for many cryptographic secure communication protocols. There are three basic categories namely-OT_1^2, OT_1^n and OT_h^n in which OT_1^2 deals with retrieving a single message from two messages privately with the guarantee that the user can only get a single message out of two messages, OT_1^n deals with retrieving a single message bit from n message bits privately with the guarantee that the user can only get a single message bit out of n message bits and OT_h^n deals with retrieving h message bits from n message bits privately with the guarantee that the user can only get h message bits out of n message bits.

Several efforts were initiated in [2–8] to make the OT_1^2 practical using Decisional Diffie-Hellman (DDH) assumption. Further, much research work [9–16] has been carried out in the construction of OT_h^n.

Motivation: Most of the existing OT schemes concentrated only on providing the highest level of privacy in presence of the underlying exponential modular operations. The basic claim is that the public key operation generally involves exponential operations (like RSA, DDH, Goldwasser-Mical, Paillier etc). If the exponential operation has been involved, then there is no chance of getting the ciphertext size less than the plaintext size. Also, the existing OT schemes does not provide inbuilt integrity support in the protocol i.e., there is no tamper evidence that can be available for the verifier (or the user).

Our solution: Conventionally, all the proposed schemes and their variants are PBR schemes by default.

We have successfully constructed a new quadratic residuosity based single database oblivious transfer scheme to achieve almost non-trivial communication and to provide a tamper evidence to the verifier. The proposed construction has the following results. (1) The new quadratic residuosity based concurrently executing recursive 2-bit encryption functions collectively support almost non-trivial communication and a tamper evidence (i.e., integrity verification) to the user. (2) The computation overhead of the proposed construction is linear (i.e., $\mathcal{O}(n)$) to the size of the database. In other words, there are no exponential modular multiplication operations involved during the response creation process.

Organization: Section 2 refers to the basic preliminaries and notations, Sect. 3 refers to the algebraic frameworks, QRA based proposed OT scheme and security proofs, Sect. 4 refers to the extensions to the basic construction, Sect. 5 refers to the performance evaluation of the proposed schemes, Sect. 6 refers to the advantages of the proposed constructions over the existing schemes, Sect. 7 refers to the real-world applications that can adopt the proposed schemes, Sect. 8 refers to the final remarks with the open problems.

2 Preliminaries and Notations

Notations: Let $[i] \triangleq \{1, 2, \ldots, i\}$ and $[i, j]$ is the process of selecting all the elements from i to j iteratively. Let $\mathcal{A}_{1\text{-to-}i} \triangleq \{\mathcal{A}_1, \mathcal{A}_2, \ldots, \mathcal{A}_i\}$. Let $N \in \{0, 1\}^k$ be the RSA modulus. Let Q_R and $\overline{Q}_R$ be the quadratic residuosity and quadratic non-residuosity sets of $\mathbb{Z}_N^{+1}$ respectively where N is the RSA composite.

Preliminaries:

Definition 1 (*A Single database oblivious transfer with almost non-trivial communication*): Let n bit 2-dimensional matrix database be $\mathcal{DB} = \{\mathcal{B}_1, \mathcal{B}_2, \ldots, \mathcal{B}_u\}$ with u rows and v columns where $|\mathcal{B}_i: i \in [u]| = v$. The proposed construction is a 3-tuple (QG,RC,RR) oblivious transfer protocol involving two communicating parties called *user* ($\mathcal{U}$) and *server* ($\mathcal{S}$). User $\mathcal{U}$ randomly generates computationally bounded query set $\mathcal{Q}_{1\text{-to-}u} = \{\mathcal{Q}_1, \mathcal{Q}_2, \ldots, \mathcal{Q}_u\}$ and sends to the server $\mathcal{S}$ using *query generation* (*QG*). Using the received query set and the database $\mathcal{DB}$, server $\mathcal{S}$ generates OT response $\mathcal{R}_{1\text{-to-}u} = \{\mathcal{R}_1, \mathcal{R}_2, \ldots, \mathcal{R}_u\}$ with almost non-trivial communication cost using *response creation* (*RC*). In-turn, user $\mathcal{U}$ retrieves the required block from the response $\mathcal{R}_{1\text{-to-}u}$ using *response retrieval* (*RR*). The protocol is a single round and single database OT with almost non-trivial communication if the following conditions hold.

- *User privacy*: $\forall i, j \in [u]$, $|P[(\mathcal{Q}_i, sk) \xleftarrow{R} QG(1^k): A(n, \mathcal{R}_{1\text{-to-}u}, 1^k) = 1] - P[(\mathcal{Q}_j,$ $sk) \xleftarrow{R} QG(1^k): A(n, \mathcal{R}_{1\text{-to-}u}, 1^k) = 1]| = neg(k) \Leftrightarrow$ QRA
- *Server privacy*: $\forall z \in [u]$, $P[RR((\mathcal{R}_{1\text{-to-}u}, sk): \mathcal{R} \xleftarrow{R} RC(\mathcal{Q}_{1\text{-to-}u}, \mathcal{DB}, n, 1^k), (\mathcal{Q}_{1\text{-to-}u},$ $pk) \xleftarrow{R} QG(1^k)) = \mathcal{B}_z] = 1$ and $\forall w \in [u]$, $w \neq z$, $P[RR((\mathcal{R}_{1\text{-to-}u}, sk): \mathcal{R}_{1\text{-to-}u} \xleftarrow{R}$ $RC(\mathcal{Q}_{1\text{-to-}u}, \mathcal{DB}, n, 1^k), (\mathcal{Q}_{1\text{-to-}u}, pk) \xleftarrow{R} QG(1^k)) = \mathcal{B}_w] = 0$
- *Correctness*: $\forall z \in [u]$, $P[RR((\mathcal{R}_{1\text{-to-}u}, sk): \mathcal{R}_{1\text{-to-}u} \leftarrow RC(\mathcal{Q}_{1\text{-to-}u}, \mathcal{DB}, n, 1^k),$ $(\mathcal{Q}_{1\text{-to-}u}, pk) \xleftarrow{R} QG(1^k)) = \mathcal{B}_z] = 1$
- *Almost Non-trivial communication*: $(\mathcal{R}_{1\text{-to-}u} \leftarrow RC(\mathcal{Q}_{1\text{-to-}u}, \mathcal{DB}, n, 1^k) : (\mathcal{Q}_{1\text{-to-}u},$ $pk) \xleftarrow{R} QG(1^k))$ such that $SIZE(\mathcal{R}_{1\text{-to-}u}) < o(n) + logN^2)$

where $P[\cdot]$ is the privacy revealing probability, (pk, sk) is the public and private key pair, A is a distinguishing server, k is the security parameter, n is the database size and $SIZE(\cdot)$ is the size function in bits.

3 Proposed Single Database Oblivious Transfer Scheme

We have introduced a novel method to achieve almost non-trivial oblivious transfer in a single database setting. All the proposed algorithms are drawn from Definition 1.

3.1 New Algebraic Frameworks

QRA based encryption: It is a new quadratic residuosity assumption based encryption method specifically designed to receive the QRA based queries and produce both QRA covered ciphertexts and almost non-trivial communication bits.

This new public key based encryption function receives two plaintext bits and produces the ciphertext. But, the decryption receives the ciphertext and the second plaintext bit (which was encrypted in encryption) and produces the first plaintext bit. For all given plaintext bits $a, b \in \{0, 1\}$ and the random input $x \in \mathbb{Z}_N^{+1}$ and the public key components $e_1 \in Q_R, e_2 \in \overline{Q}_R$, the new public key based encryption function encrypts the bits a, b and produces the ciphertext y as follows.

$$\mathcal{E}(a, b, N, x, e_1, e_2) = \begin{cases} x \cdot e_1 \cdot e_1 \equiv y \ (\text{mod } N) \text{ if } a = 0, b = 0 \\ x \cdot e_1 \cdot e_2 \equiv y \ (\text{mod } N) \text{ if } a = 0, b = 1 \\ x \cdot e_2 \cdot e_1 \equiv y \ (\text{mod } N) \text{ if } a = 1, b = 0 \\ x \cdot e_2 \cdot e_2 \equiv y \ (\text{mod } N) \text{ if } a = 1, b = 1 \end{cases} \tag{1}$$

For all the given ciphertext y and the private key components d_1, d_2 and the quadratic residuosity of the input x and the second bit b, the respective decryption function of Eq. 1 produces the first bit a as follows.

$$\mathcal{E}^{-1}(b, N, y, d_1, d_2) = \begin{cases} [y \cdot d_1 \cdot d_1 \equiv x \ (\text{mod } N) \text{ if } y \in Q_R, b = 0] \text{ implies } a = 0 \\ [y \cdot d_1 \cdot d_2 \equiv x \ (\text{mod } N) \text{ if } y \in \overline{Q}_R, b = 1] \text{ implies } a = 0 \\ [y \cdot d_2 \cdot d_1 \equiv x \ (\text{mod } N) \text{ if } y \in \overline{Q}_R, b = 0] \text{ implies } a = 1 \\ [y \cdot d_2 \cdot d_2 \equiv x \ (\text{mod } N) \text{ if } y \in Q_R, b = 1] \text{ implies } a = 1 \end{cases} \tag{2}$$

How can we get the second bit b during decryption? Instead of using a single encryption function, use the recursive encryption functions in which each successive encryption functions are again connected through the QRA based trapdoor mapping function of Freeman et.al [17] as follows.

Trapdoor based encryption: Revisit the QRA based Lossy TrapDoor Function (LTDF) of Freeman et al. [17] which receives the random inputs $x \in \mathbb{Z}_N^*$, $r \in \overline{Q}_R$ with Jacobi symbol -1, $s \in Q_R$ and produces the ciphertext y as $x^2 \cdot r^{jx} \cdot s^{hx} \equiv y$ (mod N) where $jx = 1$ if $JS(x) = -1$ otherwise $jx = 0$ and $hx = 1$ if $x > n/2$ otherwise $hx = 0$. For all $x \in \mathbb{Z}_N^{+1}$, the value of jx is always zero. If the value of hx is taken as "trapdoor" then the modified trapdoor function is

$$\mathcal{F}_{ltdf} = (x^2 \equiv y \ (\text{mod } N)) \tag{3}$$

In order to decrypt the ciphertext y, find the respective square roots using (jx, hx) combination. Note that for all ciphertext y, there are four square roots- say x_1 corresponding to $(jx = 0, hx = 0)$, x_2 corresponding to $(jx = 0, hx = 1)$, x_3 corresponding to $(jx = 1, hx = 0)$, x_4 corresponding to $(jx = 1, hx = 1)$. Since jx is always zero for all input $x \in \mathbb{Z}_N^{+1}$, the square root must be x_1 or x_2 (knowing hx is sufficient to find the square root).

Recursive Connection: For all the given plaintext bits $a, b, c, h \in \{0, 1\}$ and the random input $x \in \mathbb{Z}_N^{+1}$ and the public key components $e_1 \in Q_R, e_2 \in \overline{Q_R}$, the connective encryption function $\mathcal{C}$ for any two successive encryption functions of Eq. 1 encrypts the bits a, b, c, h and produces the ciphertext y as follows.

$$\mathcal{E}(c, h, N, (\mathcal{E}(a, b, N, x, e_1, e_2))^2, e_1, e_2) \tag{4}$$

where hx value of $\mathcal{E}(a, b, x, e_1, e_2)$ is stored as a trapdoor bit.

Consider the plaintext bit set $\mathcal{B} = \{b_i \in \{0, 1\}: 1 \le i \le v\}$. Consider the subsets with ordered pair of bits $\mathcal{D}_1 = \{(b_i, b_{i+2}): i = i + 2, i \in [2, v-2]\}, \mathcal{D}_2 = \{(b_i, b_{i+2}): i = i + 2, i \in [1, v-3]\}, \mathcal{D}_3 = \{(b_i, b_{i+1}): i = i + 2, i \in [1, v-1]\}, \mathcal{D}_4 = \{(b_i, b_{i+1}): i = i + 2, i \in [2, v-2]\}$ and group the possible pair of subsets such as $(\mathcal{D}_1, \mathcal{D}_2)$, $(\mathcal{D}_1, \mathcal{D}_3)$, $(\mathcal{D}_4, \mathcal{D}_2)$, and $(\mathcal{D}_4, \mathcal{D}_3)$ in such a way that the concatenation of the bits of that subset pair should always equal to the block $\mathcal{B}$. For instance, we have considered one of the subset pairs $(\mathcal{D}_1, \mathcal{D}_3)$ to explain the proposed scheme described below.

For all given plaintext bit subset $\mathcal{D}_3$ and random input $x \in \mathbb{Z}_N^{+1}$ and public key components $e_1 \in Q_R, e_2 \in \overline{Q_R}$, two concurrently executing recursive encryption functions are

$$\begin{aligned} (y_1, \mathcal{T}_1) &= \mathcal{E}_i(b_{v-2}, b_v, N, (\mathcal{E}_{i-1}(b_{v-4}, b_{v-2}, N, (\mathcal{E}_{i-2})^2, e_1, e_2))^2, e_1, e_2) \\ (y_2, \mathcal{T}_2) &= \mathcal{E}_j(b_{v-1}, b_v, N, (\mathcal{E}_{j-1}(b_{v-3}, b_{v-2}, N, (\mathcal{E}_{j-2})^2, e_1, e_2))^2, e_1, e_2) \end{aligned} \tag{5}$$

where $i \in [(v/2)-1, 3], j \in [v/2, 3]$ and $\mathcal{E}_{i=1}(b_1, b_2, x, e_1, e_2), \mathcal{E}_{j=1}(b_2, b_4, x, e_1, e_2)$. Also, hx values of the ciphertext of each $\mathcal{E}_i(b_{l-2}, b_l, N, \mathcal{E}_{i-1}, e_1, e_2), 1 \le i \le (v/2) - 2, 4 \le l \le v$, are stored in $\mathcal{T}_1$ as communication bits and hx values of the ciphertext of each $\mathcal{E}_j(b_{l-1}, b_l, N, \mathcal{E}_{j-1}, e_1, e_2), 1 \le j \le (v/2) - 1, 2 \le l \le v$, are stored in $\mathcal{T}_2$ as communication bits.

Note that every second bit obtained from decrypting $\mathcal{E}_i(b_{l-2}, b_l, \mathcal{E}_{i-1}, e_1, e_2)$, $1 \le i \le (v/2) - 2, 4 \le l \le v$, of Eq. 5 is same as every first bit needed to decrypt $\mathcal{E}_j(b_{l-1}, b_l, \mathcal{E}_{j-1}, e_1, e_2), 1 \le j \le (v/2) - 1, 2 \le l \le v$, of Eq. 5. Therefore, the second bit required to decrypt each $\mathcal{E}_j(b_{l-1}, b_l, \mathcal{E}_{j-1}, e_1, e_2)$ is obtained from decrypting every $\mathcal{E}_{i-1}(b_{l-2}, b_l, \mathcal{E}_{i-1}, e_1, e_2)$ as shown in Fig. 1.

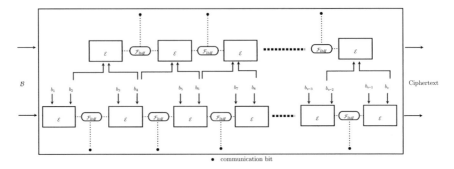

Fig. 1 Concurrently executing recursive encryption function on the block $\mathcal{B}$

3.2 Quadratic Residuosity Based Oblivious Transfer Scheme

- **Query Generation (QG)**: Let N be the RSA composite. For all $x \in \mathbb{Z}_N^{+1}$, the computationally bounded query is $\mathcal{Q}_{1\text{-to-}u} = (N, x, (e_{1,1}, e_{2,1}), \ldots, (e_{1,u}, e_{2,u}), (g, \beta))$ where $e_{1,z} \in Q_R, e_{2,z} \in \overline{Q_R}$ for all user interested block $\mathcal{B}_z, z \in [u]$, and $e_{1,w}, e_{2,w} \in Q_R$ or $e_{1,z}, e_{2,z} \in \overline{Q_R}$ for all the remaining block $\mathcal{B}_w, w \in [u]$ and $w \neq z$. Also, select $\beta \in \overline{Q_R}$ and find a generator g such that $\{\forall g \in \mathbb{Z}_{N^2}^* : (L(g^\lambda \bmod N^2))^{-1} \pmod N, L(t) = (t-1)/2, t \in \mathbb{Z}_N^*\}$ where $\lambda = lcm(p-1, q-1)$.
- **Response Creation (RC)**: Using the composite modulus N, the random input x and the public key components $e_{1,z}, e_{2,z}$ from the query $\mathcal{Q}_{1\text{-to-}u}$, each user interested block $\mathcal{B}_z, z \in [u]$, is encrypted using two concurrently executing recursive encryption functions as given in Eq. 6. Consider the subsets $\mathcal{D}_1 = \{(b_l, b_{l+2}): l = l+2, l \in [2, v-2]\}$ and $\mathcal{D}_3 = \{(b_l, b_{l+1}): l = l+2, l \in [1, v-1]\}$ of $\mathcal{B}_z$. Encrypt $\mathcal{D}_1$ using the first recursive encryption function and encrypt $\mathcal{D}_3$ using the second recursive encryption function as

$$
\begin{aligned}
(y_{1,z}, \mathcal{T}_{1,z}) &= \mathcal{E}_i(b_{v-2}, b_v, N, (\mathcal{E}_{i-1}(b_{v-4}, b_{v-2}, N, \mathcal{E}_{i-2}^2, e_1, e_2))^2, e_1, e_2) \\
(y_{2,z}, \mathcal{T}_{2,z}) &= \mathcal{E}_j(b_{v-1}, b_v, N, (\mathcal{E}_{j-1}(b_{v-3}, b_{v-2}, N, \mathcal{E}_{j-2}^2, e_1, e_2))^2, e_1, e_2)
\end{aligned} \tag{6}
$$

where $i \in [(v/2) - 1, 3]$, $j \in [v/2, 3]$ and $\mathcal{E}_{i=1}(b_1, b_2, x, e_1, e_2)$, $\mathcal{E}_{j=1}(b_2, b_4, N, x, e_1, e_2)$. Also, hx values of the ciphertext of each $\mathcal{E}_i(b_{l-2}, b_l, N, \mathcal{E}_{i-1}, e_1, e_2)$, $1 \leq i \leq (v/2) - 2$, $4 \leq l \leq v$, are stored in $\mathcal{T}_{1,z}$ as communication bits and hx values of the ciphertext of each $\mathcal{E}_j(b_{l-1}, b_l, N, \mathcal{E}_{j-1}, e_1, e_2)$, $1 \leq j \leq (v/2) - 1$, $2 \leq l \leq v$, are stored in $\mathcal{T}_{2,z}$ as communication bits. Finally, using either $y_{1,z}$ or $y_{2,z}$, encrypt the last bit $b_{z,v}$ as

$$
\mathcal{E}_{pi} = (g^{y_{1,z}} \cdot (\mathcal{E}_{gm}(\alpha, \beta, b_{z,v}))^N \equiv y_z' \pmod{N^2}) : \forall \alpha \xleftarrow{R} \mathbb{Z}_N^* \tag{7}
$$

where $\mathcal{E}_{pi}$ is the Paillier encryption and the Goldwasser-Micali encryption is $\mathcal{E}_{gm} = (\alpha^2 \cdot \beta^a \pmod N)$ such that $a = 1$ if $b_{z,v} = 1$ otherwise $a = 2$. Therefore, the block specific response $\mathcal{R}_z = (y_z', y_{2,z}, \mathcal{T}_{1,z}, \mathcal{T}_{2,z})$.

Similarly, using the composite modulus N, random input x and the public key components $e_{1,w}, e_{2,w}$ or $e_{1,w}, e_{2,w}$ from the query Q, each remaining block $\mathcal{B}_w$, $w \in [u]$ and $w \neq z$, is encrypted using two concurrently executing recursive encryption functions as given in Eq. 8. Consider the subsets $\mathcal{D}_1 = \{(b_l, b_{l+2}): l = l+2, l \in [2, v-2]\}$ and $\mathcal{D}_3 = \{(b_l, b_{l+1}): l = l+2, l \in [1, v-1]\}$ of $\mathcal{B}_w$. Encrypt $\mathcal{D}_1$ using the first recursive encryption function and encrypt $\mathcal{D}_3$ using the second recursive encryption function as

$$(y_{1,w}, \mathcal{T}_{1,w}) = \mathcal{E}_i(b_{v-2}, b_v, N, (\mathcal{E}_{i-1}(b_{v-4}, b_{v-2}, N, \mathcal{E}_{i-2}^2, e_1, e_2))^2, e_1, e_2)$$
$$(y_{2,w}, \mathcal{T}_{2,w}) = \mathcal{E}_j(b_{v-1}, b_v, N, (\mathcal{E}_{j-1}(b_{v-3}, b_{v-2}, N, \mathcal{E}_{j-2}^2, e_1, e_2))^2, e_1, e_2) \quad (8)$$

where $i \in [(v/2)\text{-}1, 3], j \in [v/2, 3]$ and $\mathcal{E}_{i=1}(b_1, b_2, N, x, e_1, e_2), \mathcal{E}_{j=1}(b_2, b_4, N, x, e_1, e_2)$. Also, hx values of the ciphertext of each $\mathcal{E}_i(b_{l-2}, b_l, N, \mathcal{E}_{i-1}, e_1, e_2), 1 \leq i \leq (v/2) - 2, 4 \leq l \leq v$, are stored in $\mathcal{T}_{1,w}$ as communication bits and hx values of the ciphertext of each $\mathcal{E}_j(b_{l-1}, b_l, N, \mathcal{E}_{j-1}, e_1, e_2), 1 \leq j \leq (v/2) - 1, 2 \leq l \leq v$, are stored in $\mathcal{T}_{2,w}$ as communication bits. Finally, using either $y_{1,w}$ or $y_{2,w}$, encrypt the last bit $b_{w,v}$ as

$$\mathcal{E}_{pi} = (g^{y_{1,w}} \cdot (\mathcal{E}_{gm}(\alpha, \beta, b_{w,v}))^N \equiv y'_w \pmod{N^2}) : \forall \alpha \xleftarrow{R} \mathbb{Z}_N^* \quad (9)$$

where $\mathcal{E}_{pi}$ is the Paillier encryption and the Goldwasser-Micali encryption is $\mathcal{E}_{gm} = (\alpha^2 \cdot \beta^a \pmod{N})$ such that $a = 1$ if $b_{z,v} = 1$ otherwise $a = 2$. Therefore, the block specific response $\mathcal{R}_w = (y'_w, y_{2,w}, \mathcal{T}_{1,w}, \mathcal{T}_{2,w})$. Therefore, the total response is $\mathcal{R}_{1\text{-to-}u} = \{\mathcal{R}_1, \mathcal{R}_2, \ldots, \mathcal{R}_u\}$ where $\mathcal{R}_l = (y'_l, y_{2,l}, \mathcal{T}_{1,l}, \mathcal{T}_{2,l}), l \in [u]$.
- **Response Retrieval (RR):** The required block $\mathcal{R}_z = (y'_z, y_{2,z}, \mathcal{T}_{1,z}, \mathcal{T}_{2,z})$ is retrieved using the respective ciphertext and communication bit set as follows.

Initially, decrypt the Paillier ciphertext y'_z of Eq. 9 to get back $y_{1,z}$ and the Goldwasser-Micali ciphertext. Then decrypt the Goldwasser-Micali ciphertext to get back the last bit $b_{z,v}$. Using the private key p, q and the respective response $y_{1,z}, y_{2,z}, \mathcal{T}_{1,z}, \mathcal{T}_{2,z}$ decrypt the recursive encryption functions to get the required block as

$$(b_{2,z}, b_{4,z}, \ldots, b_{v,z}) = \mathcal{E}_i^{-1}(b_{4,z}, \sqrt[jx=0,hx]{(\mathcal{E}_{i+1}^{-1}(b_{6,z}, \sqrt[0,hx]{\mathcal{E}_i^{-1}}, d_{1,z}, d_{2,z}), d_{1,z}, d_{2,z})}$$
$$(b_{1,z}, b_{3,z}, \ldots, b_{v,z}) = \mathcal{E}_j^{-1}(b_{2,z}, \sqrt[jx=0,hx]{(\mathcal{E}_{j+1}^{-1}(b_{4,z}, \sqrt[0,hx]{\mathcal{E}_i^{-1}}, d_{1,z}, d_{2,z}), d_{1,z}, d_{2,z})}$$
$$(10)$$

where $1 \leq i \leq (v/2) - 1, 1 \leq j \leq (v/2), e_{1,z} \cdot d_{1,z} \equiv 1 \pmod{N}, e_{2,z} \cdot d_{2,z} \equiv 1 \pmod{N}$. The notation $\sqrt[jx,hx]{y}$ indicates the quadratic root of y when jx,hx values of the input are given (refer Freeman et al. [17] for quadratic residuosity based square root calculation).

3.3 Security Proofs

User privacy: The proposed construction is based on the well-known intractability assumption called quadratic residuosity assumption. Therefore, until and unless the underlying quadratic residuosity assumption is secure, the proposed constructions always preserve the user privacy.

Server privacy: The proposed construction uses two kinds of public key components pairs in which one pair consists of two components with different quadratic residuosity properties for all user interested blocks and the other pair consists of two components with same quadratic residuosity property for all remaining blocks. It is always possible to decrypt only those ciphertexts which involve public key components with different quadratic residuosity properties. Therefore, it is always guaranteed to retrieve user interested blocks thereby ensuring the server privacy.

Integrity verification: If the residue number obtained during decryption process through RR algorithm is same as the random input sent through the query, then it is always guaranteed that the ciphertext sent from the server has not altered.

Almost Non-trivial communication: From Eqs. 6 and 8 it is evident that there are concurrently executing $(v/2) - 1$ and $v/2$ number of 2-bit encryption functions. Each concurrently executing successive encryption functions generates one communication bit in between them. Therefore, there are $(v/2) - 2$ and $(v/2) - 1$ number of communication bits generated concurrently. In total, there are $u(v - 3)$ number of communication bits (acts as communication bits from server to user) generated during *RC* algorithm.

4 Extensions and Generalizations

Multi-threaded OT scheme: In this extension, consider the subsets $\mathcal{D}_1 = \{(b_i, b_{i+2}):$ $i = i + 2, i \in [2, v - 2]\}$ and $\mathcal{D}_2 = \{(b_i, b_{i+2}): i = i + 2, i \in [1, v - 3]\}$ of any block $\mathcal{B}$. Encrypt $\mathcal{D}_1$ using the first recursive encryption function and encrypt $\mathcal{D}_2$ using the second recursive encryption function as given in Eqs. 6 and 8 for all user interested and the remaining blocks. This extended scheme is same as the proposed scheme, but only difference is the selection of block subsets and the encryption of last two bits. In this, last two bits $b_{l,v-1}, b_{l,v}$ are encrypted as $\forall l \in [u], \forall \alpha \xleftarrow{R} \mathbb{Z}_N^*$, $(g^{y_{1,l}} \cdot (\mathcal{E}_{gm}(\alpha, \beta, b_{l,v-1}))^N \equiv y'_{1,l} \pmod{N^2})$ and $(g^{y_{2,l}} \cdot (\mathcal{E}_{gm}(\alpha, \beta, b_{l,v}))^N \equiv y'_{2,l}$ $\pmod{N^2})$. The practical advantage of this scheme is that *RC* algorithm on the server side can be concurrently executed as child sub-processes since there is no bit overlap in the subsets. Similarly, *RR* algorithm on the user side can be concurrently executed as child sub-processes during block retrieval since there is no bit dependency of one process over the other. This scheme greatly reduces the overall protocol execution time since there is a vast support for concurrent process execution.

5 Performance Analysis

Communication: In the proposed OT scheme, user sends $\mathcal{O}((3k+2ku)+(\log N^2))$ bits query to the server and the server generates $mathcalO(u((v-3)+(\log N^2)))$ bits response. In the *multi-threaded* OT scheme, user sends $mathcalO((3k+2ku)+(\log N^2))$ bits query to the server. Server generates $\mathcal{O}(u((v-4)+(2\log N^2)))$ bits response.

Computation: In the proposed OT scheme, server performs $\mathcal{O}(u(3v-4))$ number of modular multiplications modulo N plus u number of Paillier encryptions and user performs $\mathcal{O}(3v-4)$ number of modular inverse multiplications modulo N plus one Paillier decryption. In the *multi-threaded* OT scheme, server performs $\mathcal{O}(u(3v-6))$ number of modular multiplications modulo N plus $2u$ number of Paillier encryptions and user performs $\mathcal{O}(3v-6)$ number of modular inverse multiplications modulo N plus two Paillier decryption.

Concurrent execution: In the proposed OT scheme, since the bits are arranged in a criss-cross manner in the subset pair $(\mathcal{D}_1,\mathcal{D}_3)$, parallel response creation for the first and second subset of the pair is possible on the server side but only sequential block retrieval is possible on the user side. In the *multi-threaded* scheme, since the bits are arranged in independent manner $(\mathcal{D}_1,\mathcal{D}_2)$, parallel response creation is possible on both user and server side.

6 Advantages

Increased security over eavesdropper by dynamic input selection: In this paper, only the input $x \in Q_R$ is considered. As an extension, the proposed scheme works for all the combinations of the input and public key components like $(x \in \overline{Q}_R, e_{1,l} \in Q_R, e_{2,l}\overline{Q}_R), (x \in \overline{Q}_R, e_{1,l} \in \overline{Q}_R, e_{2,l}Q_R), (x \in Q_R, e_{1,l} \in Q_R, e_{2,l}\overline{Q}_R), (x \in Q_R, e_{1,l} \in \overline{Q}_R, e_{2,l}Q_R)$. This dynamic random selection of query elements increases the privacy level considerably.

We consider subset pair for block: In this paper, we have selected the subset pair $\mathcal{D}_1 = \{(b_i, b_{i+2}): i = i + 2, i \in [2, n-2]\}$ and $\mathcal{D}_3 = \{(b_i, b_{i+1}): i = i + 2, i \in [1, n-1]\}$. As an extension, there are other subsets like $\mathcal{D}_2 = \{(b_i, b_{i+2}): i = i + 2, i \in [1, n-3]\}$, $\mathcal{D}_4 = \{(b_i, b_{i+1}): i = i + 2, i \in [2, n-2]\}$ and the subset combinations like $(\mathcal{D}_1, \mathcal{D}_2)$, $(\mathcal{D}_1, \mathcal{D}_3)$, $(\mathcal{D}_4, \mathcal{D}_2)$, and $(\mathcal{D}_4, \mathcal{D}_3)$. Dynamic selection of (agreed by both the parties) a subset pair among the possible pairs further increases the possibility to break the privacy of the communicating information by the eavesdropper. Further, instead of considering a subset pair, subset triple, subset quadruple etc to attain increased security concurrent process (encryption and decryption) executions.

Existing database infrastructure: The database with existing data structure can easily adopt the proposed schemes since the schemes work on a plain database. This greatly reduces the cost of database modifications.

Computationally feasible: Since the proposed schemes involve only linear number of modular multiplications, it is always feasible for practical purposes.

7   Applications

Academics: Suppose the Head of the department wants to check the progress of their faculties privately from the centralized server controlled by the Principal or vice versa. In another case, Principal wants to check the one of the CCTV footages privately. The proposed scheme effectively includes these scenarios where the Head of the department acts as the user entity and the Principal acts as the server entity in first case and Principal acts as the user entity and the CCTV operator acts as the server entity.

HealthCare: The proposed scheme can be adopted by HealthCare in which Patient, Doctor and Cloud storage entities are involved. In this, Patient (user$_1$) encrypts (*RC* algorithm) his Health record through the proposed scheme with Doctor's public key, keeps the ciphertexts with him and stores the communication bit set on the Cloud. Whenever the Patient meets the Doctor, the Doctor Obliviously downloads the Patient's communication bit record from the Cloud and gets the ciphertext of the original record from the Patient and retrieve the Patient's record using proposed decryption (i.e., *RR* algorithm). This approach preserves the record on the Cloud and preserves Doctor's privacy on the Cloud. All the entities are almost computationally divided.

8   Conclusion with Open Problems

We have successfully constructed concurrently executing recursive 2-bit encryption function as described in Eq. 5 to support both the quadratic residuosity based single database oblivious transfer with almost non-trivial communication cost and linear computation cost. We have further extended the basic construction to a more realistic multi-thread enabled oblivious scheme which reduces the overall protocol execution time. Even though the computation cost is minimized, the communication cost is linear to the database size. Therefore, attaining the logarithmic communication in intractability assumption based schemes is still an open challenge.

References

1. M.O. Rabin, in How to exchange secrets with oblivious transfer. Harvard University Technical Report 81 talr@watson.ibm.com 1295. Received 21 Jun 2005
2. A. Bill, I. Yuval, R. Omer, Priced oblivious transfer: How to sell digital goods, in *Advances in Cryptology (EUROCRYPT 2001)*, Innsbruck, Austria (2001)

3. M. Naor, B. Pinkas, Efficient oblivious transfer protocols, in *Proceedings of the Twelfth Annual ACM-SIAM Symposium on Discrete Algorithms (SODA '01)* (Society for Industrial and Applied Mathematics, 2001, pp. 448–457)
4. S. Halevi, Y.T. Kalai, Smooth projective hashing and two-message oblivious transfer. J. Cryptol. **25**(1), 158–193 (2012)
5. A.Y. Lindell, in *Efficient Fully-Simulatable Oblivious Transfer* (Springer, Berlin, 2008), pp. 52–70
6. J.A. Garay, D. Wichs, H.S. Zhou, Somewhat non-committing encryption and efficient adaptively secure oblivious transfer. Cryptology ePrint Archive, Report 2008/534 (2008) https://eprint.iacr.org/2008/534
7. Y. Aumann, Y. Lindell, in *Security against covert adversaries: Efficient protocols for realistic adversaries* (Springer, Berlin, 2007), pp. 137–156
8. C. Peikert, V. Vaikuntanathan, B. Waters, in *A framework for efficient and composable oblivious transfer* (Springer, Berlin, 2008), pp. 554–571
9. Naor, M., Pinkas, B.: Oblivious transfer and polynomial evaluation. In: Proceedings of the Thirty-first Annual ACM Symposium on Theory of Computing. STOC '99, ACM (1999) 245–254
10. M. Naor, B. Pinkas, in *Oblivious transfer with adaptive queries* (Springer, Berlin, 1999), pp. 573–590
11. M. Naor, B. Pinkas, Computationally secure oblivious transfer. J. Cryptol. **18**(1), 1–35 (2005)
12. Y. Ishai, J. Kilian, K. Nissim, E. Petrank, in *Extending oblivious transfers efficiently* (Springer, Berlin, 2003), pp. 145–161
13. J.B. ielsen, Extending oblivious transfers efficiently—how to get robustness almost for free. Cryptology (2007). ePrint Archive, Report 2007/215
14. W. Ogata, K. Kurosawa, Oblivious keyword search. J. Complex **20**(2–3), 356–371 (2004)
15. M. Green, S. Hohenberger, Blind identity-based encryption and simulatable oblivious transfer. Cryptology (2007). ePrint Archive, Report 2007/235. https://eprint.iacr.org/2007/235
16. B. Zeng, X. Tang, C. Hsu, A framework for fully-simulatable h-out-of-n oblivious transfer. CoRR (abs/1005.0043)
17. D.M. Freeman, O, Goldreich, E. Kiltz, A. Rosen, G. Segev, More constructions of lossy and correlation-secure trapdoor functions. Cryptology (2009). ePrint Archive, Report 2009/590. http://eprint.iacr.org/2009/590

Free Space Optical Communication Channel Modelling with PIN Receiver

Suman Debnath, Bishanka Brata Bhowmik and Mithun Mukherjee

Abstract Free space optical (FSO) communication is a mode of optical communication, where the data transmission channel is established via free space, rather using conventional optical fibre in optical communications. The transmission uses the free space (e.g. air) as the medium, a low-power light amplification by stimulated emission of radiation (LASER) as a transmitter and a semiconductor as the receiver. As the channels in optical fibre communication (OFC) and FSO communication are different, the losses and noises are also different in both cases. The quality of optical signal transmission through wireless depends on the atmospheric characteristics, like rain, wind, snowfall, fog, temperature, sunlight, light from other sources and turbulence. The aim of this publication is to model the channel for the optical signals through the air by considering all the losses and noises over the medium. The noises in the receiver, e.g. shot noise and thermal noise, are also analysed with on–off keying and direct detection method and have shown the effects on the output electrical signal. Bit error rate (BER) versus distance is obtained considering the above noises and losses over the channel and at the receiver. Finally, a complete FSO system is simulated by combining both the channel losses and noises at receiver.

Keywords Free space optical communication · Channel modelling · Atmospheric losses · Turbulence · On–off keying · Pin photodetector · Receiver noise

S. Debnath (✉) · B. B. Bhowmik
Department of Electronics & Communication Engineering,
Tripura University, Suryamaninagar, Tripura 799022, India
e-mail: neel.debnath1@gmail.com

B. B. Bhowmik
e-mail: bishankabhowmik@tripurauniv.in

M. Mukherjee
Guangdong Provincial Key Lab of Petrochemical Equipment Fault Diagnosis,
Guangdong University of Petrochemical Technology, Maoming, China
e-mail: m.mukherjee@ieee.org

© Springer Nature Singapore Pte Ltd. 2020
A. Elçi et al. (eds.), *Smart Computing Paradigms: New Progresses and Challenges*,
Advances in Intelligent Systems and Computing 767,
https://doi.org/10.1007/978-981-13-9680-9_22

1   Introduction

Free space optical (FSO) [1] communication or optical wireless communication (OWC) [2] is an advanced and low-cost communication technique in the modern era. This transmission overcomes several drawbacks in traditional data transmission system. For example, it omits the usage of wired channel, having high data rate (10 Gbps) with low bandwidth occupations and less link interference. This technique is a powerful alternative to radio frequency (RF) and optical fibre communication (OFC) [3].

In FSO communications, light amplification by stimulated emission of radiation (LASER) transmitters are used for high-speed communication, whereas light sources like light-emitting diodes (LEDs) are used as the transmitter for low speed (10 mbps). Typically, FSO communications operate in tera-Hertz (wavelength 800–1600 nm) unlicensed spectrum band. As a result, a huge bandwidth of light beam allows faster transmission. FSO technique widely used the line-of-sight (LOS) transmission. CAPANINA [4] is a project where a downlink transmission from a stratospheric platform of distance 60 km was established and a link length of 150 km established between two Hawaiian Islands [5]. In an indoor system, FSO can be possible with non-LOS transmission by reflecting the light from the wall to the receiver.

In addition, FSO communication provides an extreme security over data transmission as it uses a very narrow light beam and travels in a LOS path which is impossible to detect with a spectral analyser or RF meter. The receiver must be perfectly aligned to the beam, and the combination of transceiver must match with the system; then, the transmission path will be completed. In addition, FSO system (Fig. 1) exhibits the advantages of OFC systems, such as high data rate, no interference with the other electromagnetic wave like microwave, radio system and ease of installation. Moreover, the cost per bit is even lower than a traditional OFC system. On the other side of this technique, it is very much dependant on the quality of the medium, e.g. air. Bad weather degrades the transmission. The rain, wind, fog, sunlight or light from another source can affect the transmitted light beam and make it attenuated and noisy. Noises also add to the light beam due to the turbulence of air, and the loss due to it is called scintillation loss (fluctuation of the intensity). Beam wandering, beam broadening, and angle fluctuations are also caused due to turbulence [3]. As this system is LOS, objects, e.g. tree and building, in the path of the transmission also block the light beam to reach the receiver.

In this work, we study the effects of weather on the optical signal in FSO system. We have simulated the rain loss, fog loss, geometric loss and turbulence effects which cause a significant degradation to the FSO transmission quality.

In the case of the receiver, generally, two kinds of detectors are used in FSO system, PIN photodetector and avalanche photodetector (APD). These detectors also generate current noises while receiving the optical signals. The noises are shot or quantum noise, thermal or Johnson noise and dark current (exclusively in APD). So the quality of electric signal retrieved from optical signal also depends on the quality of the receiver. Here, in this work, we use a PIN photodetector.

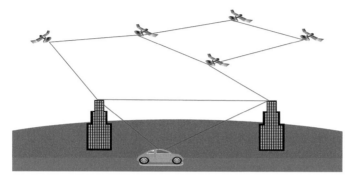

Fig. 1 A typical FSO communication system

1.1 Contribution

The objective of the paper is to study a complete FSO system combining the effects of all the losses and noises over the channel as well as at the receiver. FSO communication has been studied extensively and has simulated the noise effects on the optical signal, but we tried to combine all the atmospheric loss and noises in the receiver end and simulate the final electrical output of the whole FSO communication system. The rest of the paper is organised as follows. Section 2 discusses the losses and noises over the channel. The losses and noises at the receiver are presented in Sect. 3. The simulation results are shown in Sect. 4, and the conclusion is drawn in Sect. 5.

2 Losses Over the Channel

Losses in the atmosphere play a vital role in the degradation of the FSO transmission. The attenuation mainly occurs by scattering and absorption of light. In this work, we model the channel considering the rain loss, snow loss, geometric loss and scintillation loss.

2.1 Geometric Loss

The geometric loss is a transmission loss due to the deviation of light. This loss depends on the diameter of both transmitter and receiver and the angle of deviation.

So, more field of view (FOV) results in more geometric loss. This loss is independent of the weather conditions of atmosphere and is expressed as [6]

$$P_{\text{geo}} = -20 \log \left[\frac{d_{\text{receiver}}}{d_{\text{transmitter}} + (l \times \theta)} \right] \tag{1}$$

where P_{geo} is the geometric loss (in dB), d_{receiver} and $d_{\text{transmitter}}$ are the receiver and transmitter diameters (in m), respectively, l denotes the length of the link (in m), and θ represents the angle of beam FOV (in rad).

2.2   Turbulence and Scintillation Loss

Different studies are done, and various theoretical models have been proposed on signal degradation and intensity fluctuation due to turbulence [7–16]. The turbulence (C_n^2) occurs due to simultaneous changes in pressure, temperature and velocity in the air. Generally, the turbulence ranges from 10^{-13} to 10^{-16} m$^{-2/3}$ [17]. Due to the turbulence, the molecules distribute randomly, and as a result, the light beam has to face a fluctuation in its intensity which is called scintillation. The scintillation variance is expressed as [18]

$$\sigma_s^2 = 23.16 \ C_n^2 \ K^{7/6} \ L^{11/6} \tag{2}$$

where C_n^2 is turbulence (m$^{-2/3}$), $K = 2\pi/\lambda$ is optical wave number, and L is the link distance (in m). The above calculation is based on the spectrum of the refractive index fluctuation by Kolmogorov [19].

2.3   Fog Loss

Theoretically, the fog attenuation on light is based on Mie scattering [20]. But the popularly used two models to determine the fog loss are Kim model and Kurse model [21]. These models are based on the visibility of air through fog. The fog attenuation (in dB) is calculated as [6, 22]

$$P_{\text{fog}} = \frac{10 \log V_{\%}}{V} \left(\frac{\lambda}{\lambda_0} \right)^{-q}, \tag{3}$$

where $V_{\%}$ is percentage air drop transmission, V is visibility (in km), λ is transmitted light wavelength (in nm), λ_0 is visibility reference wavelength (in nm), and q is wavelength dependency.

The wavelength dependency expressed in both Kim and Kurse models, respectively, is as follows [6, 22, 23].

$$q = \begin{cases} 1.6 & \text{if } V > 50\,\text{km} \\ 1.3 & \text{if } 6\,\text{km} < V < 50\,\text{km} \\ 0.16V + 0.34 & \text{if } 1\,\text{km} < V < 6\,\text{km} \\ V - 0.5 & \text{if } 0.5 < V < 1\,\text{km} \\ 0 & \text{if } V < 0.5\,\text{km} \end{cases} \tag{4}$$

$$q = \begin{cases} 1.6 & \text{if } V > 50\,\text{km} \\ 1.3 & \text{if } 6\,\text{km} < V < 50\,\text{km} \\ 0.585V^{1/3} & \text{if } V < 6\,\text{km} \end{cases} \tag{5}$$

2.4 Snow Loss

Snow fall consists of two types of snows, namely dry snow and wet snow. Therefore, the snow loss is determined based on types of snow [21]. The snow loss (in dB/km) is calculated as

$$P_{\text{snow}} = a \times S^b \tag{6}$$

In the case of dry snow,

$$a = 5.42 \times 10^{-5}\lambda + 5.4958776 \quad b = 1.38 \tag{7}$$

and in the case of wet snow,

$$a = 1.023 \times 10^{-4}\lambda + 3.7855466 \quad b = 0.72, \tag{8}$$

where S is snow rate (in mm/h).

2.5 Rain Loss

Rain loss is also a significant attenuation in the FSO system. The loss (in dB) due to rain is calculated as [23]

$$P_{\text{rain}} = 1.076 R^{2/3}, \tag{9}$$

where R is rain rate (in mm/h).

3   Noises in the PIN Receiver

We have used the OOK modulation technique for simulation, so the received optical field envelop of power $P_r(t)$ can be written as

$$P_r(t) = \begin{cases} P_{t_1}(t) + a(t) & \text{for bit 1} \\ P_{t_0}(t) + a(t) & \text{for bit 0} \end{cases} \tag{10}$$

and

$$a(t) = \Delta a(t) - A \tag{11}$$

where $P_{t_1}(t)$ and $P_{t_0}(t)$ are transmitted optical power for bit 1 and bit 0, respectively, $a(t)$ is the channel noise, A is the attenuation due to rain, fog, snow and geometric loss, and $\Delta a(t)$ is scintillation noise due to turbulence.

So assuming the responsivity of the receiver is unity, the photocurrent $I_p(t)$ is [24]

$$I_p(t) = \frac{P_r(t)\eta q}{h\nu} \tag{12}$$

where η is the quantum efficiency, q is the electron charge, and $h\nu$ is the energy of proton.

3.1   Shot Noise

Shot or quantum noise develops when the photodetector converts the photons of light to photoelectron. The fluctuations in the amount of photons create a discrete flow of electron in photodetector which leads to the shot noise [25]. This noise development in photodetector follows the Poisson process [24]. The shot current noise (in Ampere) is calculated as

$$\sigma_{sn}^2 = 2qI_p(t)B \tag{13}$$

where q is the charge of electron (in Coulomb), $I_p(t)$ is receiver photocurrent (in Ampere), and B is the bandwidth of the receiver.

3.2   Thermal Noise

For all electrical circuitry, load resistance creates a noise calls thermal or Johnson noise. This noise can be reduced with a large load resistor which fulfils the requirement of receiver bandwidth [24]. The thermal noise variance (in Ampere) is described as

Table 1 Simulation parameters

Parameters	Values
Transmitted power	$-10\,$dBW
Bit rate	2.5 Gb/s
Wavelength (λ)	1550 nm
Turbulence (C_n^2)	High (10^{-13} to $10^{-14}\mathrm{m}^{-2/3}$)
Rain rate (R)	20 mm/h
Snow rate (S)	0 mm/h
Receiver diameter (d_{receiver})	13 cm
Transmitter diameter ($d_{\text{transmitter}}$)	1 mm
Angle of deviation (θ)	5 mrad
Pseudorandom bit sequence (PRBS)	$10^{10} - 1$
Visibility (V)	5 km
Visibility reference wavelength (λ_0)	550 nm
Percentage air drop transmission ($V_\%$)	5%

$$\sigma_{tn}^2 = \frac{4K_B T}{R_L} B \qquad (14)$$

where K_B is Boltzmann's constant, T is absolute temperature (K$°$), R_L is load resistance of 500Ω, and B is the receiver bandwidth of 50 GHz.

4   Simulation Results

In this work, we have studied an FSO system combining all the attenuation and noises due to atmospheric conditions and the receiver quality. A pseudorandom bit sequence of 25×2^{12} bit has been sent from the transmitter. In Fig. 2a, the total loss(dB) of the optical signal is shown up to a link distance of 5 km. From the simulation, we get that the major loss of optical power is due to geometric loss. If the system is perfectly aligned with a transmitter, having low FOV can enhance the quality of transmission. The optical power degradation against distance up to 5 km is shown in Fig. 2b.

Using the on–off keying (OOK) modulation, we calculated the bit error rate (BER) against the transmission distance from 100 m up to 5000 m in Fig. 3a. The BER values are estimated when the number of error bits was more than or equal to 100. This ensures the 95% confidence interval with ± 0.15 dB for estimating the optical signal-to-noise ratio (OSNR) [26]. The sent bit sequence from the transmitter is compared by performing XOR operation with the bit sequence of output electrical signal at the receiver to detect the number of erroneous bits. From the BER graph, it is observable that with all channel and receiver noises and attenuation in given conditions, we can retrieve the optical signal by photodetector up to link distance 800 m. But beyond the

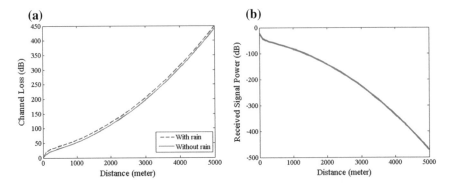

Fig. 2 **a** Loss profile of the channel, **b** received power at receiver with respect to distance when transmitted power is −10 dBW

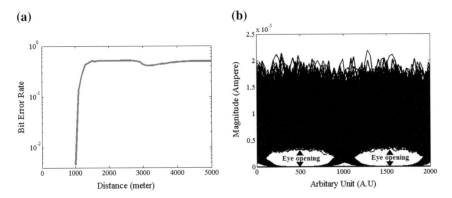

Fig. 3 **a** Performance of BER with different distance, **b** eye diagram of electrical signal when link length 500 m

distance, a major degradation of the quality of optical signal occurs. Also keeping the transmission distance of 500 m, we have taken the eye diagram of the electrical signal from the receiver in Fig. 3b. The fluctuations in the eye diagram shown are due to the turbulence in the air, and the eye-opening reduces very rapidly due to this turbulence.

In this simulation, we also observed the effects of receiver noise in the electrical signal. Omitting the channel noises in the optical signal, two eye diagrams of the electric signal at are taken—one is without the receiver noise Fig. 4a and another with receiver noise Fig. 4b to distinguish the effects of receiver noise on electrical signal at the receiver. The transmitted power and the link distance, in this case, were −10 dBmW and 10 km, respectively.

(a) **(b)**

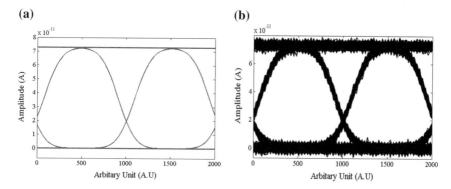

Fig. 4 **a** Eye diagram of current signal without receiver noise, **b** eye diagram of current signal with receiver noise

5 Conclusion

Atmospheric quality plays a vital role in FSO transmission. In this paper, we have analysed and combined most of the atmospheric loss and noises and modelled a transmission channel in simulation level. It is also observed that even in high turbulence the BER is very good up to transmission distance of 800 m and rain causes a significant loss of optical signal. The PIN receiver noises also took account of the simulation, and also it is observed that a good photodetector with low noises can have better sensing of the received optical signals. Overall, an FSO system has been modelled to observe the noise effects on FSO system which helps to understand and to implement this system at physical level.

References

1. S.M. Navidpour, M. Uysal, M. Kavehrad, Ber performance of free-space optical transmission with spatial diversity. IEEE Trans. Wirel. Commun. **6**(8), 2813–2819 (2007)
2. T. Fath, H. Haas, Performance comparison of mimo techniques for optical wireless communications in indoor environments. IEEE Trans. Commun. **61**(2), 733–742 (2013)
3. TRAI, Free space optics in next generation wireless networks. Technol. Dig. (8), 1–6 (2012)
4. J. Horwath, M. Knapek, B. Epple, M. Brechtelsbauer, B. Wilkerson, Broadband backhaul communication for stratospheric platforms: the stratospheric optical payload experiment (stropex). Proc. SPIE **6304**, 6304–6304 (2006)
5. D.W. Young, J.E. Sluz, J.C. Juarez, M.B. Airola, R.M. Sova, H. Hurt, M. Northcott, J. Phillips, A. McClaren, D. Driver, D. Abelson, J. Foshee, Demonstration of high-data-rate wavelength division multiplexed transmission over a 150-km free space optical link. Proc. SPIE **6578**, 6578–6578 (2007)
6. H. Kaushal, V. Jain, S. Kar, *Free Space Optical Communication* (Optical Networks, Springer India, 2017)
7. X. Zhu, J.M. Kahn, Free-space optical communication through atmospheric turbulence channels. IEEE Trans. Commun. **50**(8), 1293–1300 (2002)

8. M. Born, E. Wolf, *Principles of Optics: Electromagnetic Theory of Propagation, Interference and Diffraction of Light* (Elsevier, 2013)
9. J.W. Goodman, *Statistical Optics* (Wiley, 2015)
10. A. Ishimaru, Theory and application of wave propagation and scattering in random media. Proc. IEEE **65**(7), 1030–1061 (1977)
11. S. Karp, R.M. Gagliardi, S.E. Moran, L.B. Stotts: Optical Channels: Fibers, Clouds, Water, and the Atmosphere (Springer Science & Business Media, 2013)
12. S.K. Norman, Y.S. Arkadi Zilberman, Measured profiles of aerosols and turbulence for elevations of 2 to 20 km and consequences of widening of laser beams. Proc. SPIE **4271**, 4271–4271–9 (2001)
13. A. Zilberman, N.S. Kopeika, Y. Sorani, Laser beam widening as a function of elevation in the atmosphere for horizontal propagation. Proc. SPIE **4376**, 4376–4376–12 (2001)
14. L. Andrews, R. Phillips, *Laser Beam Propagation Through Random Media* (Press Monographs, SPIE Optical Engineering Press, 1998)
15. X. Wu, P. Liu, M. Matsumoto: A study on atmospheric turbulence effects in full-optical free-space communication systems. In: 2010 6th International Conference on Wireless Communications Networking and Mobile Computing (WiCOM) (2010), pp. 1–5
16. T. Ha, A. Duong, P. Anh, Average channel capacity of freespace optical mimo systems over atmospheric turbulence channels. ASEAN Eng. J. Part A **5**(2), 57–66 (2015)
17. I. Jacobs, C. Bean, *Fine Particles, Thin Films and Exchange Anisotropy*, vol. 3 (Academic Press Inc., New York, 1963)
18. R. Teixeira, A. Rocha, Scintillation prediction models compared with one year of measurements in aveiro, portugal. In: Antennas and Propagation Conference, 2007. LAPC 2007. Loughborough, IEEE (2007), pp. 313–316
19. P. Deng, M. Kavehrad, Z. Liu, Z. Zhou, X. Yuan, Capacity of mimo free space optical communications using multiple partially coherent beams propagation through non-kolmogorov strong turbulence. Opt. Express **21**(13), 15213–15229 (2013)
20. J.C. Maxwell, *A Treatise on Electricity and Magnetism*, vol. 2, 3rd edn. (Clarendon, Oxford, 1892)
21. A. Ishimaru, *Wave Propagation and Scattering in Random Media* (IEEE Press, An IEEE OUP classic reissue, 1997)
22. S.S. Muhammad, P. Kohldorfer, E. Leitgeb, Channel modeling for terrestrial free space optical links. In: Proceedings of 2005 7th International Conference Transparent Optical Networks, 2005, vol. 1., IEEE (2005), pp. 407–410
23. M.C.A. Naboulsi, H. Sizun, F. de Fornel, Fog attenuation prediction for optical and infrared waves. Opt. Eng. **43**, 43–43 (2004)
24. G. Keiser, *Optical Fiber Communication*, 4th edn. (TataMcGraw-Hill Education, New Delhi, 2005)
25. T. Okoshi, K. Kikuchi, *Coherent Optical Fiber Communications* (KTK Scientific Publishers, Tokyo, 1888)
26. M. Jeruchim: Techniques for estimating the bit error rate in the simulation of digital communication systems. IEEE J. Sel. Areas Commun. **2**(1), 153–170 (1984)

Authentication Methods, Cryptography and Security Analysis

Anomaly-Based Detection of System-Level Threats and Statistical Analysis

Himanshu Mishra, Ram Kumar Karsh and K. Pavani

Abstract This paper presents various parameters for the analysis of threats to any network or system. These parameters are based on the anomalous behavior of the system. To characterize the behavior of the system connected to the Internet, we need to consider a number of incoming and outgoing packets, the process running in the background and system response which include CPU utilization and RAM utilization. Dataset is collected for the above-mentioned parameter under the normal condition and under the condition of any cyber-attack or threat. Based on the deviation in the values under two conditions, another statistical parameter entropy is calculated. This will helps us to identify the type of threats.

Keywords Entropy · Threats · Anomaly detection · System response

1 Introduction

We live in the era in which world is connected to the Internet right from identity verification to communication and various other government infrastructures. The more we are connected to the Internet, the more vulnerable is our data. Secure transfer of data is of prime concern nowadays because of high-profile breaches and sophisticated cyber-attacks. The one way to curtail data exfiltration is the analysis of threats. In information security, threat refers possible hazard that can compromise the system vulnerability to gain access. Due to these threats and cyber-attacks, system shows some abnormal behavior. Characterizing this behavior and extracting some useful information out of it are called anomaly detection.

H. Mishra (✉) · R. K. Karsh · K. Pavani
National Institute of Technology, Silchar, Silchar 788010, Assam, India
e-mail: him.nits@gmail.com

R. K. Karsh
e-mail: tnramkarsh@gmail.com

K. Pavani
e-mail: psidh18@gmail.com

© Springer Nature Singapore Pte Ltd. 2020
A. Elçi et al. (eds.), *Smart Computing Paradigms: New Progresses and Challenges*,
Advances in Intelligent Systems and Computing 767,
https://doi.org/10.1007/978-981-13-9680-9_23

271

Anomaly-based intrusion detection has been the area of intense research in the past decade. There are three main classes of detection approach [1]: (a) signature-based technique—it identifies bad patterns, viz. malware; (b) anomaly-based technique—deviation of the traffic or system response from the normal pattern; it mostly uses machine learning technology; and (c) specification-based technique that is based on manually developed specifications that capture legitimate (rather than previously seen) system behaviors rather than relying on machine learning.

Artificial intelligence [2] technique is also deployed in the field of intrusion detection based on the anomaly. The radial basis function is used rather than the conventional approach to reducing the false alarm rate.

The entropy-based system is incorporated into anomaly detection technique to identify [3] multilevel distributed denial-of-service (DDOS) attack. When an attack occurs, notification is sent to the clients and cloud service provider (CPS).

Some more works are done in the field of anomaly detection [4–6]. But none of the works clearly described the method to collect dataset. Also, the old technique does not incorporate lateral file movement and back running process. So, they have limited scope to identify a novel attack.

The technique which can identify the DOS attack, DDOS attack, remote access of the system using Trojan, etc., has been explored in this paper. In DOS attack, there is flooding of the packet or we can say a false request to the victim's machine. It simply slows down the server. In DDOS attack, this request comes from different IP addresses. So, it becomes difficult for the firewall to identify such attack as a false request from different IPs seems to be a normal situation for the firewall. Therefore in order to detect it, we need to consider the entropy value of packet length. In case of DDOS attack, packet length does not vary as the payload is not changed in case of flooding.

Usually, windows defender is used to identifying Trojan. Windows defender generates an alarm as soon as any malicious file is about to run. But nowadays, various tools are available to encode such Trojan and hence cannot be recognized by any IDS. To overcome such, we need to continually monitor the process running in the background and lateral file movement as well in case it is an active attack. The main characteristics of Trojan are that it used an unused port of the system. The computer system has 65,535 ports, out of which some ports are reserved for different protocols and some are free.

This paper trammels an insight into the most practical approach to identify the characteristic of various system attacks using latest tools. Wireshark is used to capture packets flowing in the network. Process Explorer is used for collecting the data for system response. Using control panel, we can observe the process running in the background. In case of Trojan, it can easily be identified. But normal users do not always do so. For them, lateral file movement alert service can be deployed.

After the collection of the dataset, analysis has been performed using the calculation of statistical parameter entropy. At the same time, plot is drawn to recognize the type of attack.

2 Dataset Generation Method

It is the foremost preliminary step for proceeding with any intrusion detection system. This paper presents three types of dataset. Each is having its own significance in identifying the abnormalities in the network and analysis of threats.

a. **Packet flow through the system**—This packet is captured using Wireshark or any other network analyzer tool. This packet has the following information.

   (1) Source and destination IP address.
   (2) Source and destination port number.
   (3) Packet length.
   (4) Type of protocol.
   (5) Description of the packet.

These values are collected for a particular host under normal condition, and behavior of the system is analyzed. Normal packet flow is 300–400 packet/s. The packet of the different protocols is observed in the system when connected to the Internet. Also, packet length varies as payload changes in case of different protocols. Also, the source and destination IP addresses change continually.

b. **Background process and vulnerability scan**—A background process is a computer process that runs behind the scenes (i.e., in the background) and without user intervention. Typical tasks for these processes include logging, system monitoring, scheduling, and user notification. There is always a high chance of malicious process running in the background in the case when any system is compromised by a hacker. This background process is mainly some kind of Trojan. Simple Trojan can easily be identified by our system, but these are encoded nowadays.

There are various kinds of Trojan, viz. HTTP Trojan, proxy Trojan, remote access Trojan and Botnet Trojan. All kinds of Trojan occupy a free port of our system and can create a backdoor for the attacker. So, keeping a close eye on the port is necessary. Apart from these, there can be various kinds of payload that can be injected into the system depending upon the vulnerability. Firstly, our system is scanned for the vulnerability. Nessus is a tool used for security scanning remotely, which scans a computer and raises an alert if it discovers any vulnerabilities that malicious hackers could use to gain access to any computer you have connected to a network.

Few Important Information from Nessus
Apart from the list and type of vulnerability, Nessus also provides the necessary details of the system, which includes name of:

1. MAC address of the system—Nessus gives the MAC address of the system. If you want to scan the system for next time, you can match the MAC address, even if the IP address is changed.
2. Operating system—Nessus provides us the details of the OS being used along with its version, for example Windows 8.1.

3. List of the open ports—Nessus scans for each port and brings the details of services running on different ports.
4. Remedy to patch up the vulnerability—It suggests the remedy for the vulnerability to patch it up.
5. CVSS score of vulnerability—CVSS stands for Common Vulnerability Scoring System. It is a way to give numerical score to.
6. CVE number—Nessus provides the CVE number, using which the exploits or auxiliary can be searched in Metasploit PostgreSQL Database, for example CVE-2008-4250. Exploit corresponding to it is "/windows/sbm/ms08-067/netapi".
7. Exploitability—Corresponding to the CVE number, we look for the exploit in Metasploit Database. If exploit is found, the payload is created and injected into the victim's machine. Hence, the backdoor is created.
8. The family of vulnerability—It deals with the type of attack which can be performed using vulnerability. The family of vulnerability is Misc., DOS, remote, etc.

c. **System utilization dataset**—Monitoring the utilization of the CPU [7] has been widely used in the field of circuit design so that it consumes less power. In case of anomaly, CPU and processor utilization behavior deviate from regular [8].

System utilization formula
$U = 100 - (\%$ of time system is free$)$.
% time in idle task $=$ ((average time of background with no task) * 100%)/(average time of background task).
Idle task—It is the task with the lowest priority in the multitasking system.

For the purpose of analysis, we can collect the normal data of system utilization from the task manager or process explorer.

Codes can be written to store data of CPU, RAM, disk and network utilization. Once the normal behavior is identified, we can anticipate the attack if there is a drastic change in the system utilization. For example, in case of DOS attack, there is a huge amount of network utilization.

Also, we can monitor the lateral file movement in case of an active attack. Moreover, if there is any lateral file movement, it implies that the system is under attack.

3   Entropy in Threat Detection

Entropy, in general, is the measure of randomness in the environment. In the field of information theory, calculation of entropy serves as a statistical parameter in setting a threshold for characterizing the normal system behavior. Dataset is collected for a long period of time. Various random variables for a given dataset (as mentioned above) can be created by using time stamp; i.e., dataset is divided into various small subsets by using a time window. Entropy is calculated for above random variable given by

$$H(x) = -\sum Pi * \log(Pi)$$

4 Algorithm for the Threat Analysis

Steps for the analysis of system-level threats are as follows.

1. Calculate the normal range of entropy for packet flow which includes packet length and type of protocol.
2. Observe the process running in the background.
3. Scan for the used port.
4. Calculate the normal range of entropy CPU, RAM, disk and network utilization.
5. Use all the normal range of above datasets as a threshold to compare it with attack data.
6. Design a decision tree based on entropy to identify the type of attack.

5 Application of Algorithm in Threat Detection

Using the algorithm given in the paper, various system-level threats can be detected. Here are the analyses of various threats using system algorithm.

5.1 DOS and DDOS Attacks

In case of DOS attack, entropy of packet flowing through system increases. Packet length remains the same for almost every packet. System utilization increases as the number of requests is far more than normal value. Load on the system increases. But there is no extra background process running in the system.

Here, it can be observed from Fig. 1 that packet flow range has increased drastically in case of DOS attack. Normal packet range is around 300–400 packet/s, whereas in case of DOS attack it is around 6000 packet/s. It has reached a maximum of 9000 packet/s which clearly shows the flooding of packet.

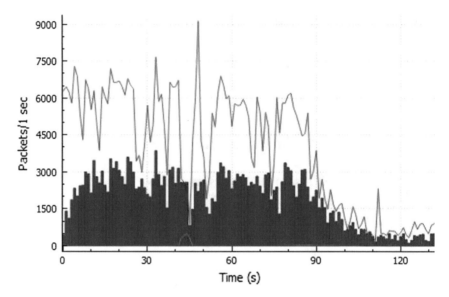

Fig. 1 Packet flow graph in case of DOS attack

5.2 Attack Using Trojan

Network utilization remains the same as usual. If an attacker tries to transfer a file from victim machine to some other machine, it can be detected by using some code which generates an alert on lateral movement of the file.

As far as the number of the incoming and outgoing packets concerns, it remains the same as the normal. These attacks cannot be identified by using the value of entropy.

5.3 ARP Spoofing

In computer networking, ARP spoofing, ARP cache poisoning or ARP poison routing are techniques by which an attacker sends (spoofed) address resolution protocol (ARP) messages onto a local area network. Generally, the aim is to associate the attacker's MAC address with the IP address of another host, such as the default gateway, causing any traffic meant for that IP address to be sent to the attacker instead. In case of ARP spoofing, entropy of a number of a packet flowing through network decreases. Each request is redirected through attacker in the middle commonly known as the man-in-the-middle attack.

Here in Fig. 2, it is shown that the number of the packets in the time 0–10 is in the normal range. But after 10 s when ARP spoofing is done, the number of the packets flowing through the network decreases. It, in turn, causes a decrease in the entropy

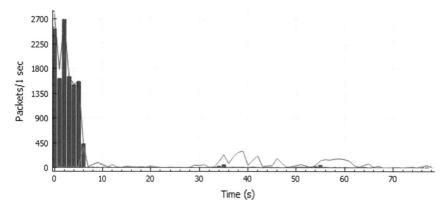

Fig. 2 Plot showing ARP spoofing

of the packet flow. Network utilization increases as there is an extra system between a user and a server.

5.4 DNS Spoofing

DNS spoofing, also referred to as DNS cache poisoning, is a form of computer security hacking in which corrupt Domain Name System data is introduced into the DNS resolver's cache, causing the name server to return an incorrect IP address. This results in traffic being diverted to the attacker's computer (or any other computer). Here in the experiment, Windows operating system is spoofed using Kali Linux.

It can be seen that the DNS client of windows machine does not forward the domain name to resolve IP to the DNS server. Instead of that, it sends the request to Kali Linux machine (attacker's machine) and shows the result which is hosted on the Web server of Kali Linux.

Figure 3 shows a Wireshark output of the system in which system could not able to connect to the configured DNS server. Message of router solicitation indicates that the router is not able to reach to the configured DNS.

6 Comparison of Result

This paper considers almost every parameter simultaneously which can be used to identify and characterize any threats to the system. So far, no work has been done in this field of analysis of threats in which a number of the packet flow, protocol type, packet length, system utilization (RAM, CPU, network and processor) and process running in the background are taken into account simultaneously as per our knowl-

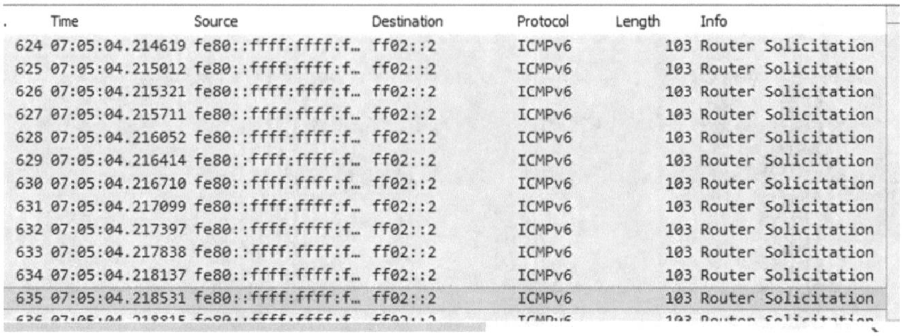

.	Time	Source	Destination	Protocol	Length	Info
624	07:05:04.214619	fe80::ffff:ffff:f…	ff02::2	ICMPv6	103	Router Solicitation
625	07:05:04.215012	fe80::ffff:ffff:f…	ff02::2	ICMPv6	103	Router Solicitation
626	07:05:04.215321	fe80::ffff:ffff:f…	ff02::2	ICMPv6	103	Router Solicitation
627	07:05:04.215711	fe80::ffff:ffff:f…	ff02::2	ICMPv6	103	Router Solicitation
628	07:05:04.216052	fe80::ffff:ffff:f…	ff02::2	ICMPv6	103	Router Solicitation
629	07:05:04.216414	fe80::ffff:ffff:f…	ff02::2	ICMPv6	103	Router Solicitation
630	07:05:04.216710	fe80::ffff:ffff:f…	ff02::2	ICMPv6	103	Router Solicitation
631	07:05:04.217099	fe80::ffff:ffff:f…	ff02::2	ICMPv6	103	Router Solicitation
632	07:05:04.217397	fe80::ffff:ffff:f…	ff02::2	ICMPv6	103	Router Solicitation
633	07:05:04.217838	fe80::ffff:ffff:f…	ff02::2	ICMPv6	103	Router Solicitation
634	07:05:04.218137	fe80::ffff:ffff:f…	ff02::2	ICMPv6	103	Router Solicitation
635	07:05:04.218531	fe80::ffff:ffff:f…	ff02::2	ICMPv6	103	Router Solicitation
636	07:05:04.218915	fe80::ffff:ffff:f…	ff02::2	ICMPv6	103	Router Solicitation

Fig. 3 Plot of the number of packets per second in DNS attack

edge. Also, statistical parameter entropy is used as a threshold for characterizing the normal behavior of the system. The normal behavior of any system depends on its configuration; therefore, the paper presents the relative change in the value of the parameter.

7 Conclusion

All the threats analyzed here are the few common threats. But cyber-crime is increasing at an alarming rate, so cybersecurity should go hand in hand. Identification and analysis of novel attack require thorough investigation. So using the algorithm, discussed in the paper it becomes an easy procedure to identify novel threats.

A little variation in the algorithm can be used to identify network-level threats as well. There is also the wide scope of research in the area of application-level threats. A penetration test is also a way to find some novel vulnerability. If it is found, then the patch can be released before the zero-day attack.

References

1. R. Sekar, A. Gupta, J. Frullo, T. Shanbhag, A. Tiwari, H. Yang, S. Zhou, Specification-based anomaly detection: a new approach for detecting network intrusions, in *CCS'02: Proceedings of the 9th ACM Conference on Computer and Communications Security* (2002)
2. R. Ravinder Reddy, Network intrusion anomaly detection using radial basis function networks. Int. J. Res. Comput. Sci. 1011–1014 (2017)
3. A.S. Navaz, V. Sangeetha, C. Prabhadevi, Entropy based anomaly detection system to prevent DDoS attacks in cloud. arXiv preprint arXiv 1308–6745 (2013)
4. M. Tavallaee, N. Stakhanova, A.A. Ghorbani, Toward credible evaluation of anomaly-based intrusion-detection methods. IEEE Trans. Syst. Man Cybern. Part C (Appl. Rev.) **40**(5), 516–524 (2010)
5. V. Chandola, A. Banerjee, V. Kumar, Anomaly detection: a survey. ACM Comput. Surv. (CSUR) **41**(3), 15 (2009)

6. F. Sabahi, A. Movaghar, Intrusion detection: a survey, in *ICSNC'08, 3rd International Conference on IEEE Systems and Networks Communications*, 23–26 Oct 2008
7. S.T. Kung, C.C. Cheng, C.C. Liu, Y.C. Chen, Dynamic power saving by monitoring CPU utilization. U.S. Patent, 574,739, Jun 2003
8. R.K. Shymasundar, N.V. Narendra Kumar, P. Teltumde, Realizing software vault on Android through information-flow control, in *2017 IEEE Symposium on Computers and Communications (ISCC)* (2017), pp. 1007–1014

A New Modified Version of Standard RSA Cryptography Algorithm

Sudhansu Bala Das, Sugyan Kumar Mishra and Anup Kumar Sahu

Abstract Data security is an open research challenge due to the rapid evolution in communication over an unsecured network. In this direction, cryptographic algorithms play an important role in providing the data security against malicious attacks. This paper elaborates a new asymmetric cryptography approach, i.e. modified RSA algorithm. This proposed method reduces complex calculation involved in *RSA algorithm*. A novel algorithm has been proposed for finding the value of public key (k) and private key (l). Also, a newly derived formula is used to find the ciphertext (N) and plaintext (M). We have explored some major research challenges in the field of *RSA algorithm*.

Keywords Encryption · Decryption · Key · Asymmetric cryptography · RSA algorithm

1 Introduction

In the context of a communication network, the use of Internet, intranet and extranet has become common and unavoidable for various basic purposes that incorporate exchange of information, data transaction, access through cloud computing and so on. Be it social media or e-banking, personal data over the network needs to be secured which otherwise may be misused by unauthorized users/hackers. Therefore, security and integrity of data over the Internet during data transactions pose a major challenge so as to minimize the risk of data pilferage and information hacking as well. In order to address this, among various methods, cryptography has been adopted as an important method since ancient time.

S. B. Das (✉) · A. K. Sahu
Defence Research and Development Organization, Balasore, Odisha, India
e-mail: baladas.sudhansu@gmail.com

S. K. Mishra
C.V. Raman College of Engineering, Bhubaneswar, Odisha, India
e-mail: sugyan3@gmail.com

© Springer Nature Singapore Pte Ltd. 2020
A. Elçi et al. (eds.), *Smart Computing Paradigms: New Progresses and Challenges*,
Advances in Intelligent Systems and Computing 767,
https://doi.org/10.1007/978-981-13-9680-9_24

281

Cryptography, a process in combination with encryption and decryption, is a practice of techniques that protect the original message by converting the same into unusable form *en route* during communication [1]. *Encryption algorithm* converts the original message (plaintext) into the unreadable format (ciphertext) by the help of the key. The output of encryption algorithm is ciphertext, whereas decryption algorithm converts the ciphertext into plaintext by the help of key [2–5]. The main goal of our paper is to reduce complex calculation in *RSA algorithm*. We have presented a new methodology for finding the value of k and l. The following paper contents are as follows. Section 2 gives an overview of a symmetric and asymmetric algorithm for encryption and decryption process. Section 3 describes our proposed methodology for finding the value of k and l. We also bring out the formula for finding ciphertext and plaintext. In Sect. 4, we have presented a graphical comparison between our proposed algorithm ($M * k$) with *RSA algorithm* (M^k). Finally, Sect. 5 concludes with the scope of future work.

2 Related Work

This section gives a summary of different types of cryptography along with its advantages and disadvantages. Cryptography algorithms are broadly classified into two types, namely symmetric-key and asymmetric-key algorithms.

Symmetric-Key cryptography is used for both encryption and decryption. It is quite simple and easy to implement. But, it has some major drawbacks. Once the single key is known, the attacker can easily find out the information [3]. In asymmetric-key cryptography, two different keys are used for encryption and decryption. The public key is distributed to all the senders, and private key is only known to a particular user (receiver). But the major drawback of this key cryptography is slower as compared to symmetric algorithms [6, 7].

Before our proposed approach, the conventional approach using *RSA algorithm* [3] is briefed. This cryptography is used as k for encryption and l for decryption by *RSA algorithm,* as follows.

Step 1. Select two large distinct prime numbers such as r and s.
Step 2. Compute $z = rs$ and $\phi(z) = (r - 1) * (s - 1)$.
Step 3. Choose a number l which is prime to $\phi(z)$, such that gcd(1,$\phi(z)$)=1, $1 \leq l \leq \phi(z)$.
Step 4. Find k such that $kl \bmod \phi(z) = 1$.

Here, *RSA algorithm* is illustrated through a simple example. In this example, $r = 3$ and $s = 11$ (*Step 1*), yielding $z=r * s=33$ and $\phi(z) =(3 - 1)*(11 - 1)= 20$ (*Step 2*). The private key l is chosen as 7, which is relatively prime to 20 (*Step 3*). The number k is derived from 7 $k \bmod 20$ =1 as $k = 3$ (*Step 4*). We have found the l value is 7 and k is 3 by using *RSA algorithm*.

According to our observation, key generation in *RSA algorithm* is very slow as it does many complex calculations that generate very large numbers. Hence, a simple formula is introduced for the calculation of private and public key. We have also found a new formula for ciphertext, i.e. $N = M * k \ mod \ z$ and for plaintext is $M = N * l \ mod \ z$. In the next section, we have illustrated our proposed algorithm through block diagram and examples.

3 Our Proposed Approach

In our approach, the plaintext is encrypted through the encryption algorithm with the help of public key. Then, the output of encryption algorithm, i.e. ciphertext is transferred over the unsecured channel to the receiver. At the receiving end, the ciphertext is converted into an original message through the decryption algorithm with the help of private key.

The proposed approach is explained through a block diagram which is shown in Fig. 1.

We have proposed a simplified method to find the value of k and l for encryption and decryption, respectively, which is described below.

Step 1. Choose two prime numbers, i.e. r and s.
Step 2. Compute $z = (r + s)$ and $\phi(z)=(r - 1) * (s - 1)$.
Step 3. Choose a number l which is relative to $\phi(z)$, such that $gcd(l,\phi(z))=1, 1 \leqslant l \leqslant \phi(z)$.
Step 4. Find k such that $kl \ mod \ \phi(z)=1$.

N is obtained as $N = M * k \ mod \ z$.
N may then be decrypted as, i.e. M,

$$
M = \begin{cases} N * l \ mod \ z + F, & \text{if } N * l \ mod \ z \leq M \\ N * l \ mod \ z - F, & \text{if } N * l \ mod \ z > M \end{cases}
$$

Let us assume the value of $r = 11$ and $s = 3$ (*Step 1*). Then, $z = 11 + 3 = 14$ and $\phi(z) = 20$ (*Step 2*). Now, we find the value of l for decryption, i.e. $gcd(l, 20) = 1$.

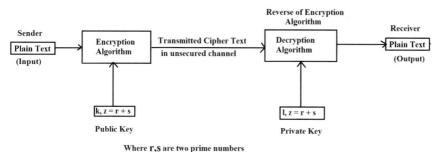

Fig. 1 Block diagram of our proposed approach

The least integer primary number l as 7 (*Step 3*). Again, we have to calculate the k for encryption, i.e. $7k \bmod 20 = 1$, so $k = 3$ (*Step 4*). In the next section, we have compared our proposed approach with *RSA algorithm*.

4 Experiment and Evaluation

In this section, we have compared the algorithms in terms of M^k and $M * k$. A formula is presented to find the value of z, N and M to avoid complex calculation in *RSA algorithm*. The experiments are performed on a single processor with *4 GB* random access memory (RAM) and *1:80 GHz* processor. Our proposed algorithm has set of inputs such as two prime numbers, and the outputs are k and l. We have found the value of k is 3 and l is 7. With these, we are able to find the ciphertext (N) and plaintext (M). Formula for finding ciphertext (N) and plaintext (M) by using *RSA algorithm* is given below.

$$N = M^k \bmod z \tag{1}$$

$$M = N^l \bmod z \tag{2}$$

Two different cases have been taken for comparison of our proposed algorithm with *RSA algorithm*.

Case 1: We have assumed our plaintext is *SIMPLE*. We have worked out encryption and decryption by using *RSA algorithm* as shown in Table 1 and also determined encryption and decryption by using our proposed approach as shown in Table 2. We have compared M^k and $M * k$ of word *SIMPLE* by using our proposed approach and *RSA algorithm* as shown in Fig. 2.

Case 2: This time, we have taken our plaintext (M) to be *SORT*. We have computed encryption and decryption by using *RSA algorithm* as shown in Table 3 and also calculated encryption and decryption by using our proposed approach as shown in Table 4. We have compared M^k and $M * k$ of word *SORT* by using our proposed

Table 1 Public-key cryptography by RSA algorithm

	M	M^k	$M^k \bmod z$	N^l	$N^l \bmod z$	M
S	19	6859	28	13492928512	19	S
I	9	729	3	2187	9	I
M	13	2197	19	893871739	13	M
P	16	4096	4	16384	16	P
L	12	1728	12	353831808	12	L
E	5	125	26	8031810176	5	E

Table 2 Proposed public-key cryptography

	M	$M * k$	$M * k$ $mod\ z$	$N * l$	$N * l$ $mod\ z$	$F = |(N * l\ mod\ z) - P|$	M
S	19	57	1	7	7	12	19 (S)
I	9	27	27	189	24	15	9 (I)
M	13	39	11	77	7	6	13 (M)
P	16	48	6	42	0	16	16 (P)
L	12	36	8	56	0	12	12 (L)
E	5	15	1	7	7	2	5 (E)

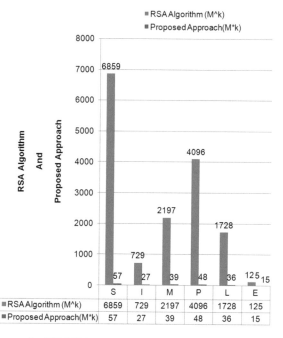

Fig. 2 Comparison result of SIMPLE word

approach and *RSA algorithm* as shown in Fig. 3. Two cases with plaintexts as SIMPLE and SORT have been performed incorporating both RSA and our proposed method whose results are compared graphically as shown in Figs. 2 and 3, respectively. It is evident that our proposed method is very effective in terms of complexity in comparison with *RSA algorithm*.

Table 3 Public-key cryptography by RSA algorithm

	M	M^k	$M^k \bmod z$	N^l	$N^l \bmod z$	M
S	19	6859	28	13492928512	19	S
O	15	3375	9	4782969	15	0
R	18	5832	24	4586471424	18	R
T	20	8000	14	105413504	20	T

Table 4 Proposed public-key cryptography

	M	$M*k$	$M*k \bmod z$	$N*l$	$N*l \bmod z$	$F = \lvert(N*l \bmod z) - M\rvert$	M
S	19	57	1	7	7	12	19 (S)
O	15	45	3	21	7	8	15 (0)
R	18	54	12	84	0	18	18 (R)
T	20	60	4	28	0	20	20 (T)

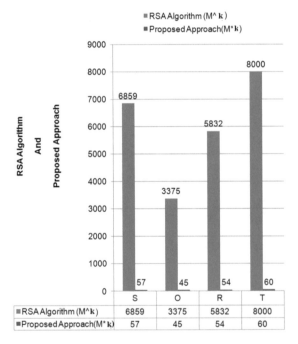

	S	O	R	T
■ RSA Algorithm (M^k)	6859	3375	5832	8000
■ Proposed Approach(M*k)	57	45	54	60

Fig. 3 Comparison result of SORT word

5 Conclusion

In today's world, the Internet is an essential component for different purposes. It is a major issue to provide secure data communication over an unsecured network. This paper presents a modified version of the RSA algorithm which used public-key cryptography technology. The major limitation of the RSA algorithm is time-consuming in the generation of the key. So, we have presented a new methodology to find the value of public and private key for encryption and decryption. A new mathematical formula has been proposed for finding the ciphertext (N) and plaintext (M). This formula helps us to reduce in the complex calculation of finding the key, ciphertext (N) and original message. We have explained our proposed approach through block diagram and examples. Also, the comparison result is shown in the form of a graph. Our future work includes design of an online algorithm for encryption and decryption process which will be helpful towards the real-life problem.

References

1. A. Kumar, R. Sharma, A secure image steganography based on rsa algorithm and hash-lsb technique. Int. J. Adv. Res. Comput. Sci. Softw. Eng. **3**, 363–372 (2013)
2. S. Kumari, A research paper on cryptography encryption and compression techniques. Int. J. Eng. Comput. Sci. **6**, 20915–20919 (2017)
3. A. Kahate, Crypt. Netw. Secur. **6**, 1–480 (2008) (Tata McGraw-Hill)
4. A. Karki, A comparative analysis of public key cryptography. Int. J. Modern Comput. Sci. **4**, 30–35 (2016)
5. P. Jovanovic, Analysis and design of symmetric cryptographic algorithms. Doctoral thesis from Computer Science and Mathematics of the University of Passau (2015), pp. 1–216
6. W. Diffie, M.E. Hellman, New directions in cryptography. IEEE Trans. Inf. Theor. **6**, 30–35 (1976)
7. A.B. Forouzan, *Data Communications and Networking* (McGraw-Hill, 2012)

Internet, Web Technology, Web Security, and IoT

e-Learning Web Services and Their Composition Strategy in SOA

Pankaj Kumar Keserwani, Shefalika Ghosh Samaddar and Praveen Kumar

Abstract With the advent of advanced communication technologies, the concept of e-learning came into practice. Most of the e-learning systems are operating in client server and distributed system environment. The activity of e-learning remained confined to predefined need due to the lack of suitable dynamic software architecture of newer technology. With the advent of service-oriented architecture (SOA), a suitable and better solution framework has been proposed for e-learning. Different functions of e-learning system are implemented as stand-alone web services. The strength of the approach followed in this paper is reusability and interoperability. To achieve the new functionality in e-learning system, the composition of relevant stand-alone web services are required to make the system fully capable of interactive services making electronic teaching learning feasible. Web service composition is a SOA-based model to make a composite web service by existing stand-alone or other composite services. In this paper, we present a full-fledged e-learning system with the help of web services and their composition as per SOA standard.

Keywords Service-oriented architecture (SOA) · Web service · e-Learning · Composite web service · Moodle

P. K. Keserwani (✉) · S. G. Samaddar · P. Kumar
Department of Computer Science and Engineering,
National Institute of Technology Sikkim, Sikkim 737139, India
e-mail: pankaj.keserwani@gmail.com
URL: http://www.nitsikkim.ac.in

S. G. Samaddar
e-mail: shefalika99@yahoo.com

P. Kumar
e-mail: pkjust4u@gmail.com

© Springer Nature Singapore Pte Ltd. 2020
A. Elçi et al. (eds.), *Smart Computing Paradigms: New Progresses and Challenges*,
Advances in Intelligent Systems and Computing 767,
https://doi.org/10.1007/978-981-13-9680-9_25

1 Introduction

Nowadays, e-learning is the most required service in the universities as well as in organizations to provide the training to their employees at their own place and at any time of the users' choice. From the developer point of view, the structure of these e-learning systems is very complex. Previously, e-learning systems were client server based; then, it shifted to distributed environment, and now, with the advent of SOA, these are shifting towards service-oriented architecture (SOA). There is a further development of cloud services in single cloud and multicloud; however, the scope of the paper allows to proble composition of web services in SOA. In SOA-based e-learning system, each functionality is implemented as stand-alone web service, and their compositions provide improved capacity of the system [1]. But composition of web services is still a problem because still there is no common standard due to multi-vendor service incompatibility. Web service composition is a very important service-orientated principle and is also known as service assemblies. Composition provides new functionalities by creating new services from existing and distributed services all over the web. Basically, web services are software platform-independent applications that export a description of their functionalities and make it available using standard technologies. Web services communicate using open protocols and are self-contained as well as self-describing. By using web services, one can plug in and immediately use new components as easily as if the functionalities are available in software libraries. SOAP, UDDI and WSDL are the basic elements of the web service implementation. It can be discovered using UDDI [2, 3]. UDDI (Universal Description, Discovery and Integration) is basically a search engine for web services and is itself a web service [4]. Web service gives the opportunity to the execution of business on distributed network and can be accessed through standard interfaces and protocols. Web service composition provides new functionalities, developed by combining pre-exiting services. But the challenging issue is the composition of these services to provide new functionality. There are many approaches for serving such purpose. Some provide semi-automatic tool [5]. In this paper, we present a SOA-based e-learning web services composition implemented using ASP.NET, as an example. These services are registered in the local Universal Description, Discovery and Integration (UDDI) as per publish–subscribe scheme of SOA. The composition of these services is done using the address and the availability of useful services from previously configured UDDI [6].

2 Background Study and Related Literature Survey

With the introduction of cloud services, there are a number of service delivery models of e-education system rather than an e-learning system. But for theoretical and research-related activities, e-learning mechanism yields fruitful results. In fact, the point happens to be one of our motivations to build up composition of web services in e-learning. Sbitneva et al. [7] followed an ontosemiotic approach for exploration into applications of theoretical constructions and methods to analyse the learning process in the e-modality of linear algebra course. The activities are arranged to make deeper comprehension for the subject. Pattnayaka and Pattnaik [8] made an attempt for the integration of web services in order to build a knowledge society. Web 3.0 has been used for such addition to knowledge base. Group work, question and answer sessions have been incorporated as web 3.0 is capable of applying artificial intelligence (AI) technique speeding up teaching learning. However, all such efforts cannot go without a baseline of security. A robust e-learning service on Moodle platform is GakuNinMoodle. The service delivery model was developed and designed with adequate security measure in Japan by Ueda and Nakamura [9]. In fact, the paper promotes security awareness education. These recent applications of learning management systems (LMSs) in Moodle platform and generation of models for test and train showed a renewed interest in the application of web services in e-learning. Systems based on web services are mainly used in the field of e-commerce. Travel and tourism systems and banking systems are some of them. Client server-based LTSA framework like stand-alone service components does not own any responsibility towards offering themselves for preparation of composite services. The composition compatibility of services only comes as an afterthought in this architecture making it a limited service architecture. Learning content provider (server) may not be able to resolve concurrency. Replication of provider due to different customizations may make data inconsistent. A full-fledged e-learning-based web services and their composition are yet to come. To support web service composition, there are a number of models and frameworks each having their pros and cons. IEEE LTSA is a very suitable model for such purpose, because it has already been extended according to the e-learning web service requirement [10]. Composite services may be accessed through a registry under prevalent publish–subscribe scheme. Types of e-learning include synchronous and asynchronous e-learning mechanism. The extended IEEE LTSA's features are as follows:

- e-learning web services reside on different locations (on different servers).
- These web services are registered in UDDI (web services location can be found at UDDI).
- A client can use these web services through the UDDI and can call these functions anywhere to make an e-learning system with the help of these services even though these services are at different physical locations.
- In this model, provider/operating server is at application layer performing services composition as well as service compatibility.

SOA-based extensible platform [11] is based on model-driven architecture (MDA) and SOA. MDA is categorized into three abstraction levels such as CIM, PIM and PSM. Computation-independent model (CIM) contains detailed information about business use case, business structure and business processes. Platform-independent model (PIM) consists of analysis and design which specifies services in technology-independent manner. Platform specification model (PSM) is used to generate the main part of the code.

Service bus provides the necessary communication infrastructure. It also provides centralized management for services. Service bus platform on SOA contains business modelling processes. It uses modelling tools and business process-based development platform to design and construct information management system. A framework for the semantic composition of web services handling user constraints [12] may be defined as the process that specifies how to select, combine and execute a set of available web services to obtain a composite service which fulfils the users' requests. The framework is based on distributed architecture, on semantic uniform ontology and on community service descriptions. Community services with the same ontology are considered as an integrator of services.

Business Process Execution Language (BPEL) provides a mechanism for formalism of the system. Web service specification is represented using UML in the form of message sequence charts and transformed into a finite-state process (FSP). Labelled Transition System Analyser (LTSA) is used to check if a web service composition satisfies the specification indicating inconsistency, if any. However, most of these related formalisms (BPEL, UML, FSP and LTSA) have been independently defined, and there is no clear connection between them [13].

3   e-Learning Web Services Publish and Subscription Methodology

e-learning system composed of a number of web services; some of them are primary web services and some are secondary. But both primary services and secondary services are equally important. The difference is brought out in their use and functionality. Primary services are those which are used for the coordination of other services as UDDI service and compatibility checking service. Secondary services include fully functional stand-alone services such as registration service, online examination service, virtual classes. Figure 1 presents IEEE LTSA model of e-learning in SOA environment [10]. In e-learning system, local UDDI is actually configured UDDI for service in Microsoft Web Server 2008. Registration service, computer programming service (lectures in text form), computer programming service (lectures in audio/video form) and virtual classes are implemented on different systems as services then registered with the service provider's name to UDDI by publish service of UDDI. UDDI service requires providers name and the services WSDL file for its publication in UDDI registry. The process is shown in Figs. 3 and 4 (Fig. 2).

The client system searches for required services and finds their location through UDDI which has already been configured. Searching e-learning web services in UDDI is typically bound on the selection of keywords pertaining to the web services, e.g. identity management, authentication, digital certificate, etc. Web services can be searched in UDDI by their keywords as explained above. Search function of UDDI returns the service provider name, service location with other details, which consists of WSDL detail as shown in Fig. 5.

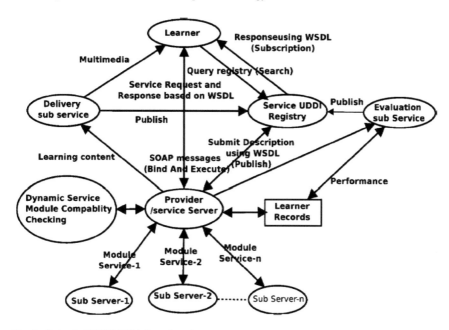

Fig. 1 Extended IEEE LTSA for e-learning system

Fig. 2 Framework for the e-learning platform-based SOA and MDA [11]

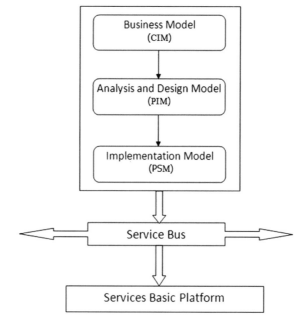

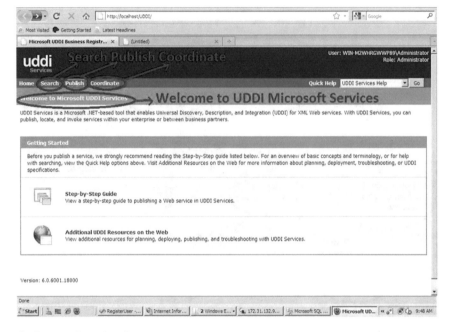

Fig. 3 UDDI home interface

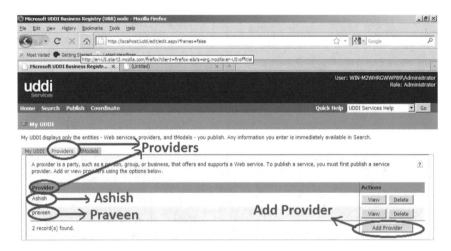

Fig. 4 UDDI provider interface

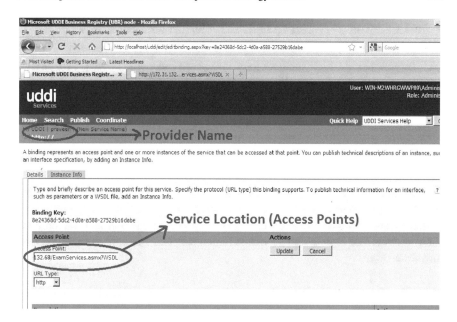

Fig. 5 UDDI add service interface

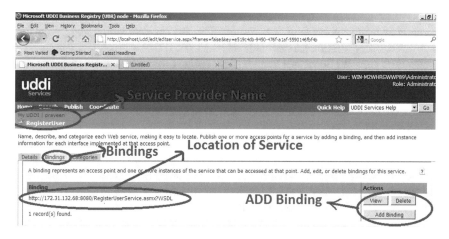

Fig. 6 UDDI add service interface

The service can now be identified in Fig. 6. UDDI service binding interface found at the searched location and is ready for use.

4 e-Learning Web Services Composition Methodology

Web service composition methodology is worked out with the help of a limited number of services of e-learning. Four types of e-learning services are offered to users:

- Computer programming service (lectures in text format)
- Computer programming service (lectures in audio/video format)
- Online examination service
- Virtual classes.

Each of these services is stand-alone and is totally independent service from others. Besides these services, there is one more service called registration service. This is also a stand-alone and independent service but provides authorized and authenticated service. The registration service is required to be composed with the services mentioned above.

4.1 Online Examination Service Composition with Registration Service

As soon as user selects the online examination service, user is directed to registration service. Registration service validates the user. If the user is already registered, registration service allows user to go back to online examination service. Otherwise, if user is not already registered, then it is forwarded to registration process. After getting successful registration, user comes back to online examination service. As soon as the user completes the examination, result of the examination gets displayed on the screen (Fig. 7).

At server side, as server gets the request for online examination service, it forwards user for his registration service, if user is registered, then server again redirects the user to online examination service where he is allowed to write the examination. If the user is not registered, then it invokes the registration service and composed it with online examination service (Figs. 8 and 9).

To show the method of composition of services, we are making use of two services: Registration service and online examination service. The method of composition is represented by the following pseudocode:

Composition Method I:

```
1) begin
2) Get service choice from user
    a) Save choice into a variable var
    b) Call Login Service with an argument of saved variable (var)
```

```
c) Do Login (validation)
   if user is (valid) then
    i) check variable (var)
    ii) call service which refer to (var)
    iii) Get the result of service
    iv) Return result
   else
    Return error to caller of Login Function
3) end
```

Composition Method II: The user selects more than two services and wants to use services in sequence. Let the user select computer programming and online examination service. Next, he uses login or registration service to access these in a sequence. If the first service (computer programming) is failed, then the second service is not called by the server and an error message will be displayed to user by frontend service. If user is not able to access, then, to learn computer programming, he is considered unprepared for writing the examination of computer programming. The pseudocode of second scenario is represented as follows:

Data Structures: An e-learning composition represents a class of services. It also uses a flag. The flag represents whether the service is successfully called or not; true value represents service is called successfully; and in case of any error, it is evaluated as false.

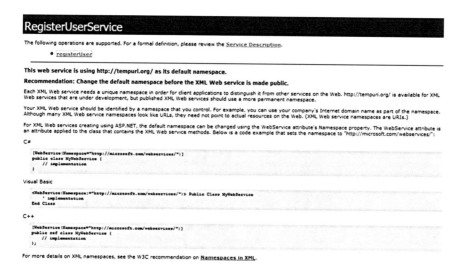

Fig. 7 UDDI service binding interface

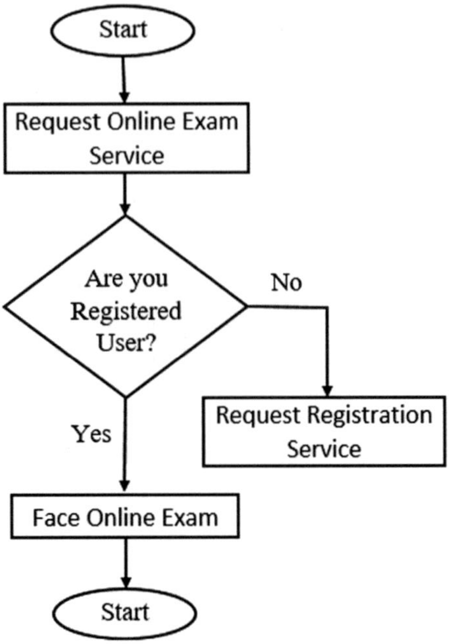

Fig. 8 Users dealing with
the system

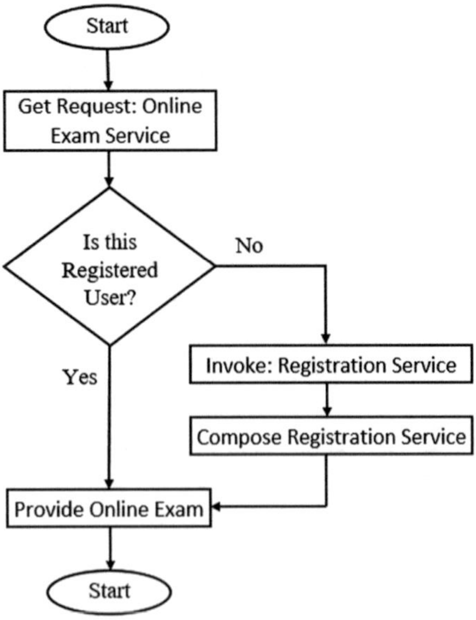

Fig. 9 Providers dealing
with the system

Steps in the method:

```
1) begin
2) Initialize e-learning composition object with null values to all
   members and false to flag
3) Get service choice from user
 a) Save choices into a variable var
 b) Call Login Service with an argument of saved  variable (var)
 c) Do Login (validation)
if user is valid then
 i) check variable (var)
 ii) call first service which refer to (var)
 iii) if (first service successful)
     A) Save its output into e-learning composition object.
     B) Set flag to true
 iv) If( flag is true )
     A) Call second service refer to (var)
     B) if(second service is successful)
         Save its output into e-learning composition object.
         Set flag to true
     C) else
         Set flag to false
     else Return error to caller of Login Function
4) end
```

5 Conclusion and Future Direction of Work

The extended IEEE LTSA e-learning framework in SOA shows immense possibility for various services composition; even on a cross-platform, cross-service provider scenario may be adopted. Integration of a number of services is possible only on Semantic Web. This paper provides a glimpse of service composition of two or three services in ASP.NET that are compatible. The concept of forward and backward compatibility would deal with the composition in a fine-grained manner that has been decided as the future direction of research. The deployment of concept requires further extension of the present model in Semantic Web-based model. With the introduction of cloud services, services composition assumes a new assemblage with third-party agency service. There are a number of cloud agents or brokers providing full or part a service. Such mechanism demands a probe in future direction of work.

Acknowledgements The authors express deep sense of gratitude to Param Kanchenjunga High Performance Computing Centre, National Institute of Technology Sikkim, Sikkim, India, where the work has been carried out.

References

1. Dion hinchliffe's SOA blog: How can we best make "the writeable web" a responsible place? microservices expo. http://soa.sys-con.com/node/173822. Accessed 30 Oct 2017
2. Seekda technology to create greater demand and greater margins. https://seekda.com/en/. Accessed 30 Oct 2017
3. XML web services. https://www.w3schools.com/xml/xml_services.asp. Accessed 30 Oct 2017
4. Recent changes, UDDI. http://uddi.xml.org/recent_changes. Accessed 30 Oct 2017
5. T. Bultan, X. Fu, R. Hull, J. Su, Conversation specification: a new approach to design and analysis of e-service composition, in *Proceedings of the 12th International Conference on World Wide Web* (ACM, 2003), pp. 403–410
6. D. Berardi, *Automatic Service Composition: Models, Techniques and Tools* (2005)
7. L. Sbitneva, N. Moreno, D. Rivera, R. Valdez, On systemic perspective and the strategies in e-learning: inquiries in linear algebra. Procedia Soc. Behav. Sci. **228**, 278–284 (2016)
8. J. Pattnayak, S. Pattnaik, Integration of web services with e-learning for knowledge society. Procedia Comput. Sci. **92**, 155–160 (2016)
9. H. Ueda, M. Nakamura, Gakuninmoodle: toward robust e-learning services using moodle in Japan. Procedia Comput. Sci. **96**, 1710–1719 (2016)
10. P. Kumar, S.G. Samaddar, A.B. Samaddar, A.K. Misra, Extending IEEEE ITSA e-learning framework in secured SOA environment, in *2010 2nd International Conference on Education Technology and Computer (ICETC)*, vol. 2. (IEEE, 2010), p. 136
11. Z. Cheng, T. Huang, J. Nong, An extensible development platform for SOA-based e-learning system, in *2008 International Conference on Computer Science and Software Engineering*, vol. 5. (IEEE, 2008), pp. 901–904
12. Y. Gamha, N. Bennacer, G. Naquet, B. Ayeb, L.B. Romdhane, A framework for the semantic composition of web services handling user constraints," in *IEEE International Conference on Web Services, ICWS'08*, vol. 2008. (IEEE, 2008), pp. 228–237
13. F. Latreche, F. Belala, A semantic framework for analyzing web services composition. Int. J. Comput. Appl. **5**(4), 47–53 (2010)

An IoT-Based Smart Plant Monitoring System

S. V. Athawale, Mitali Solanki, Arati Sapkal, Ananya Gawande
and Sayali Chaudhari

Abstract In this modern era of fast-moving technology, we can do things which we could never do before and to do these tasks there is a necessity to build a platform to perform these tasks. The proposed system puts forth the home automation technique for smart plant watering system. A smartphone empowers the user to be updated with their current garden status using IoT from any part of the world. The system uses Node ESP8266 as the microcontroller interfacing unit.

Keywords Smart plant watering · MQTT · IoT · Cloud · Soil moisture · Wi-Fi

1 Introduction

Early stages of home automation began with labor-saving machines. Home automation is the way that allows us to network our devices together and provide us immaculate control over all aspects of home. There is an increasing need of automation of tasks in our daily lives. Today, people are looking at ways and means to better their lifestyles using the latest technologies as it becomes inevitable to have easy and convenient methods and means to control and operate these appliances.

One of the aspects that this automation focuses is an efficient smart plant watering system. Plants bring a different perspective to our lives and have beneficial aesthetic properties. However, maintenance is an important task for plants as too much or too less water can damage and cause harm to the life of plants. For instance when the nursery owner or a garden owner is out of town or away from his garden, they fail to take care of the plants. The proposed system aims at providing a regularized supply of water, and status or alerts can be accessed by users from any remote location.

S. V. Athawale (✉) · M. Solanki · A. Sapkal · A. Gawande · S. Chaudhari
Department of Computer Engineering, AISSMS College of Engineering, Pune, India
e-mail: svathawale@aissmscoe.com

© Springer Nature Singapore Pte Ltd. 2020
A. Elçi et al. (eds.), *Smart Computing Paradigms: New Progresses and Challenges*,
Advances in Intelligent Systems and Computing 767,
https://doi.org/10.1007/978-981-13-9680-9_26

2 Literature Review

In OO design for an IoT-based automated plant watering system [1], the system ensures that plants are watered at a regular interval, with appropriate amount, whenever they are in need. And the system explains that soil moisture sensor checks the moisture level in the soil. Then if moisture level is below a certain limit, a signal is sent to the pump to start pumping water to plants and sets timer to required watering duration. In this paper [2], soil moisture sensor uses the principle of inverse relation between soil moisture and soil resistance. If the resistance is more, then moisture is less. In this system, Bluetooth is used for communication and it is not used for longer distances so this is the drawback. This paper [3] introduces smart irrigation system. It allows interfacing the farming and the communications and control systems. This system uses GSM module for sensing of data and sends it to database server. SMSs are sent to the mobile phone; hence, this system is not smart and used only for short ranges.

In wireless sensor and actuator system for smart irrigation on the cloud [4], cloud-based systems are used. System proposes a cloud-based wireless sensor and actuator network communication system. It monitors and controls a set of sensors, to access plant water needs. The communication protocol used in this system is Zigbee.

In automated water usage monitoring system [5], concept of the Internet of things is used to monitor and track the water usage via various sensor nodes. Wi-Fi or LAN is used to send data to LabVIEW model. LabVIEW software acts as a server to control and monitor data. Server collects the data through Wi-Fi/LAN to process and track usage and wastage of water at every outlet. The user can continuously keep a track of the water usage or wastage through a mobile with an Internet connection. In paper [6], it describes the design and implementation of an automatic water sprinkler system for a farmland which monitors and controls a water sprinkler. This system reduces waste of water and saves time. This paper [7] introduces IoT crop-field monitoring using sensors like soil moisture, humidity, light and temperature which automates the irrigation system. This system is used in greenhouses, and power consumption is reduced.

In agricultural crop monitoring using IoT—a study [8], IoT technology helps in collecting information about conditions like weather, moisture, temperature and fertility. Crop online monitoring enables detection of weed, level of water, pest detection, animal intrusion into the field, crop growth, agriculture. This paper [9] introduces a framework called Agritech. It consists of smart devices, wireless sensor network and Internet. Smart mobile phones are used to automate the agricultural processes. This paper [10] describes smart wireless home security system and home automation system. This system uses microcontroller TI-CC3200 Launchpad board and onboard Wi-Fi shield. In a low-cost smart irrigation system using MQTT protocol [11], it tries to design a simple water pump controller using soil moisture sensor and ESP8266 NodeMCU-12E. A Message Queue Telemetry Transport (MQTT) protocol is used for transmitting and receiving sensor information. A mobile application is developed, and the soil moisture sensor data and water pump status are displayed on a mobile application.

3 Proposed System

The main objective of the proposed system is to implement and design a cost-effective smart home automated system. It mainly is an implementation of IoT for remotely controlling home application of watering the plants. The system includes both hardware and software in which the hardware is the embedded system and software includes an Android application as well as a dashboard. A low-cost Wi-Fi module ESP8266 is used. Both soil-related and environment-related sensors are used. Soil sensors like moisture sensors continuously monitor the moisture in the soil. If the moisture level in the soil is less, the water pump is started. Ultrasonic distance sensors monitor the water level in the water tank. The data collected by these sensors is sent to the server. The server uses approximate values on which the system runs. A protocol named Message Queue Telemetry Transport (MQTT) is used for establishing a connection in unreliable networks. It helps in connecting the system to the application even when the user is in a remote location.

The system works as follows. The soil moisture level is checked by the sensors along with the ultrasonic sensor checking the water level in the water tank. These sensors are connected to two different NodeMCUs. These NodeMCUs collect information and send that data to the cloud server. Various analyses are performed on the data, and an appropriate value is sent to the user. If the moisture level is less, it is notified to the user and upon instruction one of the NodeMCUs starts the 5 V relay which in turn starts the system and provides water to the plants.

3.1 Software and Hardware Interface

3.1.1 Software Components

Arduino IDE: It is open-source Arduino Software (IDE).
It contains the text editor for writing a code and connects to the hardware to upload the programs and communicate with them.

3.1.2 Hardware Components

ESP8266 NodeMCU: On-chip integration with sensors and other applications.
ESP 8266 is a Wi-Fi microchip.
It has microcontroller capability and TCP/IP stack.
It has 16 GPIO pins.
2CH relay: Two-channel relay is used for controlling higher current load.
2CH relay board is a 5 V two-channel relay interface board.
Various appliances and gadgets with large current are controlled by two-channel relay module.

Ultrasonic distance sensor is used for measuring water level in the tank.
Ultrasonic sensor is a sensor which is used to sense or measure the distance of an object by using sound waves.
These are self-contained solid-state devices which are designed for non-contact sensing of solid and liquid objects.
They are used for monitoring the level of water in a tank.
Soil moisture sensor is used to read the moisture content values from the soil.
Moisture sensor senses the water level in the soil.
It estimates volumetric water content from the soil.

4 Methodology

The sensors check for the moisture level in the soil. If the moisture is found out to be deficient, then the water tank level is checked. If that amount is satisfactory, then the relay is turned on which further will turn on the fountain pump. The fountain pump will help water the plants and notify the user accordingly. If the water level in the tank is less, then user will be notified to refill the tank. Once the soil moisture reaches the expected level, the system will stop. If the soil moisture level is not satisfied, the fountain will be turned on again and plants will be watered till the desirable moisture level is reached.

4.1 Formulas

Displayed equation represents the threshold determined for the soil moisture and water level status,

$$T_M = M_C + W_L \tag{1}$$

T_M Threshold moisture level value which is required for soil.
M_C Current moisture level of soil.
W_L Water level required to reach threshold value.

If the threshold falls below a certain level, then the user will be alerted to check the water level in tank and start the pump.

5 Results

In this section, we represent the sensor data in the form of graphs. The first graph represents values of soil moisture as well as water level of present systems. The second graph represents the values of sensors of the system proposed in this paper.

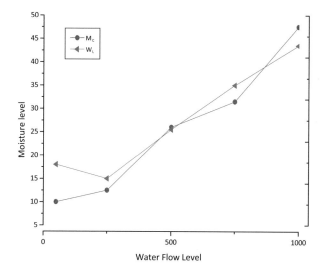

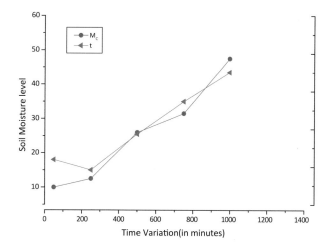

The systems which are present in Fig. 1 show inconsistency in levels of water as compared to the moisture, as plants require the appropriate water taking moisture in consideration. The system presented in this paper (Fig. 2) makes use of continuous moisture levels with respect to time, thereby providing the most appropriate levels of water. Our system will reduce the inconsistency present in the moisture and water flow level.

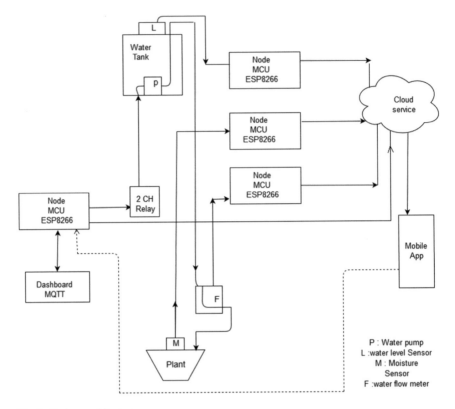

Fig. 1 System architecture

6 Conclusion and Further Direction

Upon understanding how automation works as well as its benefits and issues, it is quite interesting to look forward to how the project can be implemented in real world. The Internet of things is going to be a major part of our lives in the upcoming years. In the literature review, various features in plants were monitored using technologies like GSM, Zigbee and Bluetooth, which also have drawbacks. The proposed system has tried to overcome these drawbacks. This system when integrated into smart home automation system will be very useful for the users who will be able to monitor their system from remote location. The system asks the user to take action and gives the control to them.

In the future work, the system can be integrated to take timely actions on its own based on the predictions. The existing system can be implemented for larger datasets of large-scale applications like agriculture and big farms.

Acknowledgements The authors would like to thank the reviewers for their suggestions and remarks for improving the paper content. We also extend our gratitude to the guide and mentor Prof. S. V. Athawale for sharing his wisdom and providing an insight that assisted the project.

Fig. 2 Flowchart of the system

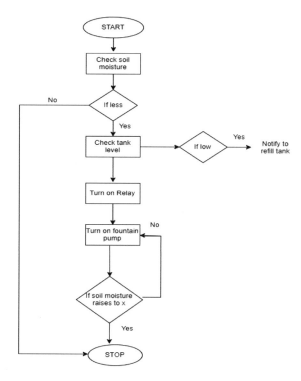

References

1. S. Rajagopal, V. Krishnamurthy, OO design for an IoT based automated plant watering system, in *IEEE International Conference on Computer, Communication, and Signal Processing* (2017)
2. A. Na, W. Isaac, S. Varshney, E. Khan, An IoT based system for remote monitoring of soil characteristics, in *2016 International Conference on Information Technology* (2016)
3. A. Byishimo, A.A. Garba, *Designing a Farmer Interface for Smart Irrigation in Developing Countries* (ACM, 2016)
4. N. Sales, O. Remdios, A. Arsenio, *Wireless Sensor and Actuator System for Smart Irrigation on the Cloud* (IEEE, 2015)
5. S. Saseendran, V. Nithya, Automated water usage monitoring system, in *International Conference on Communication and Signal Processing*, India, 6–8 Apr 2016
6. D.T. Ale, E.O. Ogunti, O. Daniela, Development of a smart irrigation system. Int. J. Sci. Eng. Invest. **4**(45) (2015)
7. P. Rajalakshmi, S. Devi Mahalakshmi, *IOT Based Crop-Field Monitoring and Irrigation Automation* (2016)
8. D.K. Sreekantha, A.M. Kavya, Agricultural crop monitoring using IOT—a study, in *2017 11th International Conference on Intelligent Systems and Control (ISCO)* (2017)
9. A. Giri, S. Dutta, S. Neogy, Enabling agricultural automation to optimize utilization of water, fertilizer and insecticides by implementing Internet of Things (IoT), in *2016 International Conference on Information Technology (InCITe)* (2016)

10. R.K. Kodali, V. Jain, S. Bose, L. Boppana, IoT based smart security and home automation system, in *International Conference on Computing, Communication and Automation (ICCCA 2016)* (2016)
11. R.K. Kodali, B.S. Sarjerao, A low cost smart irrigation system using MQTT protocol, in *2017 IEEE Region 10 Symposium (TENSYMP)* (2017)

A Study of Fog Computing Technology Serving Internet of Things (IoT)

Sumanta Banerjee, Shyamapada Mukherjee and Biswajit Purkayastha

Abstract In some applications of IoT, it is imperative to hasten the processing of data in order to get real-time conclusions. With this objective, CISCO devised fog computing, a next-generation networking technology. It resembles to the cloud but in a miniature form, which serves the edge of the network from a closer geographic location. It has the ability to enhance the yields of ubiquitous smart networks and is a step forward toward a smarter world. In this paper, an effort has been made to expound the underlying concept of fog computing technology with a comparison with other similar technologies. The challenges of this technology highlighted along with recent work progress to address them are also discussed.

Keywords Internet of Things · Fog computing · Cloud computing · Mist computing

1 Introduction

The twenty-first century is the era of Internet. It supplies highly useful information to the sectors like health care, education, business, power, etc. This treasure of information is extracted by processing huge amounts of data, produced in these sectors. The use of pervasive networks of smart objects (or things) to get information and control over the environment, is proliferating in recent years. These networks that connects smart objects (other than computers or laptops) by following a standardized protocol suit to establish communication among them, disregarding their physical lo-

S. Banerjee (✉) · S. Mukherjee · B. Purkayastha
Department of Computer Science and Engineering,
National Institute of Technology Silchar, Silchar, Assam, India
e-mail: sumanta.banerjee85@gmail.com

S. Mukherjee
e-mail: shyamamukherji@gmail.com

B. Purkayastha
e-mail: biswajitpurkayastha1960@gmail.com

© Springer Nature Singapore Pte Ltd. 2020
A. Elçi et al. (eds.), *Smart Computing Paradigms: New Progresses and Challenges*,
Advances in Intelligent Systems and Computing 767,
https://doi.org/10.1007/978-981-13-9680-9_27

311

Fig. 1 A general
architecture of fog
computing

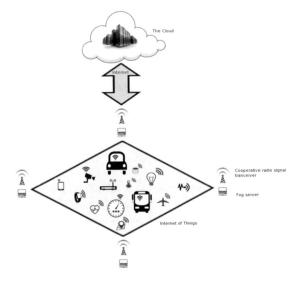

cations can be termed as Internet of Things (IoT) [1]. These networks (of wearables, routers, surveillance cameras, airline data recorders and other sensors) produce huge amounts of data every day [2]. For better decision making, these data are sent to the cloud which runs analytics on them. The extracted insights are used to command the concerned actuators.

The fog computing technology carries out processing task with transient data storage, nearer the edge of the network. A general view of a fog computing architecture is depicted in Fig. 1. It provides the necessary storage, processing, communication and management of prioritized data, produced by the sensors. It leverages decentralized processing that essentially addresses the overhead of delay in giving a response to the actuators [3]. It serves applications that require real-time response and the cloud with knowledge extracted from the data. The cloud does further processing on the processed and filtered data by fog and sends back wisdom to the fog for better discernment [4]. It exploits the cloud and edge resources and implements better co-ordination among them to achieve minimal latency, high throughput, and scalability [7].

In sectors like health care, real-time response based on the data supplied by the monitoring devices is coveted. Hazardous circumstances can be detected or predicted and can be taken care of expeditiously using fog computing approach. Some other areas like smart vehicle control, smart traffic control, real-time gaming and augmented reality that require low latency computation are its fields of application. It outperforms the cloud (in latency sensitive areas) by, (a) saving the time of transmitting huge amount of data generated by the IoT devices over long distances, (b) reducing the susceptibility of data from attacks and (c) reducing the distance which the actuation command travels.

2 Comparison of Fog with Other Similar Technologies

There are modern paradigms offering services remotely present distributed resources through the next-generation networking technologies. This section gives a concise comparison among some of such paradigms and establishes the position of fog in the picture (Fig. 2).

Cloud computing is a large set of resources that provides platform (PaaS), infrastructure(IaaS) and software(SaaS) as a service in a pay as you use fashion. It stores and processes data in remote data centers and clustered server computers that take significant time to produce insights. It is unsuitable for some systems that need faster response. **Mobile cloud computing (MCC)** technique is a cloud service for mobile devices. Mobile devices can offload their processing task to the cloud servers and conserve their resources like energy, processing and storage [4]. The **Mobile cloudlet** is formed among devices within a cluster. A device, rich with a specific resource type, may become the server for the others, whereas another device may serve the network with some other resource types, and so on. It saves the offloading time and offers a good QoS to mobile device users connected to it [5, 6]. The **mist computing** technology performs the computations within the sensors and actuators. It sends information to the server as and when the server asks for it. It saves energy (of the devices) and bandwidth (of the channel) by reducing data transmission [8].

Fig. 2 Depiction of cloud, MCC, cloudlet, fog and mist

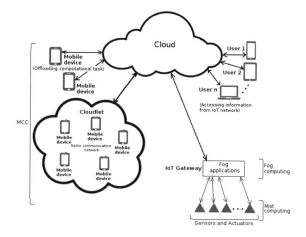

3 Challenges and Recent Works

Fog is an emerging technology to process IoT data with the aim to serve the requirements of the proliferating smart mobile devices and is still maturing. Several researchers and corporate houses are engaged in its enhancement and advancement. Here is a discussion on the challenges involved, and some recent research works addressing them are presented.

3.1 Resource Management at Real Time

Efficient management of resources to achieve minimum latency and maximum throughput with least energy and bandwidth consumption is required. The real-time fall detection algorithm [9] based on accelerometer data uses the edge devices and the server to execute the task. This [10] offers Stack4Things, a framework spanning the IaaS and PaaS layers, for Smart City Cyber Physical Systems solutions. The algorithm [11] utilizes the available fog resources and the cloud resources to offer a resource management and task scheduling scheme, to the large-scale offloading applications. The work in [12] pushes programs dynamically to devices for pre-processing the data. It achieves a congestion-free faster network by using end devices, edge networks and data centers. The MMOG [13] proposes a cloud-fog architecture to render game videos and process other user requests in some devices other than the users' device.

3.2 Performance Issues

It is challenging to extract good performance with constrained resources available in IoT networks, like to attain quick data aggregation, quick resource provisioning and node mobility. In [3], an e-Health solution, Fog CAMA is implemented and the improvements compared to the Cloud CAMA are shown. The paper [14] has proposed a dynamic video stream processing scheme for urban traffic surveillance near the capturing device (drone camera). In [15] this architecture the overloaded cloudlets offload tasks to fog servers installed in busses to cater a better QoE to the mobile users. This paper [16] employs a smart provisioning of limited resources using fog in identification and resolution of individuals by their biometric ID. It [17] proposes a load balancing algorithm for a cloud-fog-based architecture, emphasizing on uniform distribution of the load across closer and more number of edge nodes. It [18] collects clinical speech data through a smart watch and performs processing and storage work in between the sensor and the cloud and sends the processed data to the cloud storage. It [19] has proposed a fog-based smart-grid model and explained its benefits like low latency, improved security and reliability.

3.3 Network Management

Network management is also an issue for this bandwidth constrained networks. Incorporation of modern technologies such as software-defined network (SDN) [19] can enhance control of the network. This [20] algorithm implements multiuser clustering of radio access points in centralized and decentralized layers and presents a comparison. It [21] presents a bandwidth prediction model, a load balancing approach in fog infrastructures and a dynamic edge node selection mechanism for better QoS. This research [22] presents fog radio access networks (F-RAN) to decrease the heavy burden of large-scale radio signal processing in the centralized baseband unit pool.

3.4 Security and Privacy Issues

It is a challenging task to implement security and privacy in fog networks that involves resource constrained, heterogeneous and distributed nodes connected through multiple service providers. The public key infrastructure (PKI)-based solutions that involves multicast authentication is a potentially effective solution [23]. This paper [24] explains (a) network security, (b) data security, (c) access control, (d) privacy of data and location, (e) attackers interest in private data.

3.5 Fault Tolerance

Another important aspect is the fault tolerance. It is also a challenge that requires the system to be designed in a way such that it remains immune to failure of some of its components. The time-triggered distributed architecture [25] provides a fault-tolerant and real-time system. It instantaneously relocates a software component from a failed hardware to a working one.

There are some other imperative issues such as **scalability**, **platform dependence** and **interoperability**.

4 Conclusion

This study of fog computing technology focuses on its merits and significances in the emerging network architectures. It also compares the fog computing technology with other similar technologies to pinpoint its importance and efficacy in the discussed field. It can be concluded that the fog computing is an emerging technology that efficiently addresses the latency sensitivity issue of IoT networks. It is a recent approach and, hence, involves several unattended challenges that require attention of researchers and developers.

Acknowledgements Thanks to TEQIP-III for funding this paper.

References

1. S. Madakam, R. Ramaswamy, S. Tripathi, Internet of things (IoT): a literature review. J. Comput. Commun. **3**, 164–173 (2015)
2. S. Sarkary, S. Chatterjee, S. Misra, Assessment of the suitability of Fog computing in the context of Internet of Things. IEEE Trans. Cloud Comput. https://doi.org/10.1109/TCC.2015.2485206
3. F. Ramalho, A. Neto, K. Santos, J.B. Filho, N. Agoulmine, Enhancing eHealth smart applications: a Fog-enabled approach, in *2015 17th International Conference on E-health Networking, Application & Services* (*HealthCom*) (Boston, MA, 2015), pp. 323–328
4. P. Gupta, S. Gupta, Mobile cloud computing: the future of cloud. Int. J. Adv. Res. Electr. Electron. Instrum. Eng. **1**(3) (2012)
5. A.E.H.G. El-Barbary, L.A.A. El-Sayed, H.H. Aly, M.N. El-Derini, A cloudlet architecture using mobile devices, in *2015 IEEE/ACS 12th International Conference of Computer Systems and Applications* (*AICCSA*) (Marrakech, 2015), pp. 1–8
6. M. Chen, Y. Hao, Y. Li, C.F. Lai, D. Wu, On the computation offloading at ad hoc cloudlet: architecture and service modes. IEEE Commun. Mag. **53**(6), 18–24 (2015)
7. F. Bonomi, R. Milito, J. Zhu, S. Addepalli, Fog computing and its role in the Internet of Things, in *Proceedings of the First Edition of the MCC Workshop on Mobile Cloud Computing* (*MCC'12*) (ACM, New York, NY, USA, 2012), pp. 13–16
8. http://www.thinnect.com/static/2016/08/cloud-fog-mist-computing-062216.pdf
9. Y. Cao, S. Chen, P. Hou, D. Brown, FAST: a Fog computing assisted distributed analytics system to monitor fall for stroke mitigation, in *2015 IEEE International Conference on Networking, Architecture and Storage* (*NAS*) (Boston, MA, 2015), pp. 2–11
10. D. Bruneo et al., Stack4Things as a fog computing platform for Smart City applications, in *2016 IEEE Conference on Computer Communications Workshops* (*INFOCOM WKSHPS*) (San Francisco, CA, 2016), pp. 848–853
11. X.-Q. Pham, E.-N. Huh, Towards task scheduling in a cloud-Fog computing system, in *2016 18th Asia-Pacific Network Operations and Management Symposium* (*APNOMS*) (Kanazawa, 2016), pp. 1–4
12. H.J. Hong, P.H. Tsai, C.H. Hsu, Dynamic module deployment in a Fog computing platform, in *2016 18th Asia-Pacific Network Operations and Management Symposium* (*APNOMS*) (Kanazawa, 2016), pp. 1–6
13. Y. Lin, H. Shen, Cloud Fog: towards high quality of experience in cloud gaming, in *2015 44th International Conference on Parallel Processing* (Beijing, 2015), pp. 500–509
14. N. Chen, Y. Chen, Y. You, H. Ling, P. Liang, R. Zimmermann, Dynamic urban surveillance video stream processing using Fog computing, in *2016 IEEE Second International Conference on Multimedia Big Data* (*BigMM*) (Taipei, 2016), pp. 105–112
15. D. Ye, M. Wu, S. Tang, R. Yu, Scalable Fog computing with service offloading in bus networks, in *2016 IEEE 3rd International Conference on Cyber Security and Cloud Computing* (*CSCloud*) (Beijing, 2016), pp. 247–251
16. P. Hu, H. Ning, T. Qiu, Y. Zhang, X. Luo, Fog computing based face identification and resolution scheme in Internet of Things. IEEE Trans. Ind. Inform. **13**(4), 1910–1920 (2017)
17. S. Verma, A.K. Yadav, D. Motwani, R.S. Raw, H.K. Singh, An efficient data replication and load balancing technique for Fog computing environment, in *2016 3rd International Conference on*

Computing for Sustainable Global Development (*INDIACom*) (New Delhi, 2016), pp. 2888–2895

18. A. Monteiro, H. Dubey, L. Mahler, Q. Yang, K. Mankodiya, Fit: a Fog computing device for speech tele-treatments, in *2016 IEEE International Conference on Smart Computing* (*SMART-COMP*) (St. Louis, MO, 2016), pp. 1–3

19. F.Y. Okay, S. Ozdemir, A Fog computing based smart grid model, in *2016 International Symposium on Networks, Computers and Communications* (*ISNCC*) (Yasmine Hammamet, 2016), pp. 1–6

20. J. Oueis, E.C. Strinati, S. Barbarossa, Distributed mobile cloud computing: a multi-user clustering solution, in *2016 IEEE International Conference on Communications* (*ICC*) (Kuala Lumpur, 2016), pp. 1–6

21. C.F. Lai, D.Y. Song, R.H. Hwang, Y.X. Lai, A QoS-aware streaming service over fog computing infrastructures, in *2016 Digital Media Industry & Academic Forum* (*DMIAF*) (Santorini, 2016), pp. 94–98

22. M. Peng, S. Yan, K. Zhang, C. Wang, Fog-computing-based radio access networks: issues and challenges. IEEE Netw. **30**(4), 46–53 (2016)

23. S Nagar, S Setia, A detail review on Fog computing security. IJIRCCE **4**(6) (2016) https://doi.org/10.15680/IJIRCCE.2016.04063

24. P. Kumar, N. Zaidi, T. Choudhury, Fog computing: common security issues and proposed countermeasures, in *2016 International Conference System Modeling & Advancement in Research Trends* (*SMART*) (Moradabad, 2016), pp. 311–315

25. H. Kopetz, S. Poledna, In-vehicle real-time Fog computing, in *2016 46th Annual IEEE/IFIP International Conference on Dependable Systems and Networks Workshop* (*DSN-W*) (Toulouse, 2016), pp. 162–167

26. A.V. Dastjerdi, R. Buyya, Fog computing: helping the Internet of Things realize its potential. Computer **49**(8), 112–116 (2016)

Author Index

© Springer Nature Singapore Pte Ltd. 2020
A. Elçi et al. (eds.), *Smart Computing Paradigms: New Progresses and Challenges*,
Advances in Intelligent Systems and Computing 767,
https://doi.org/10.1007/978-981-13-9680-9

Printed in the United States
By Bookmasters